FRONTIERS OF SCIENCE

SPACE AND ASTRONOMY

FRONTIERS ☲ SCIENCE

SPACE AND ASTRONOMY

Notable Research and Discoveries

KYLE KIRKLAND, PH.D.

Facts On File
An imprint of Infobase Publishing

SPACE AND ASTRONOMY: Notable Research and Discoveries

Facts On File, Inc.
An imprint of Infobase Publishing
132 West 31st Street
New York NY 10001

Library of Congress Cataloging-in-Publication Data

Kirkland, Kyle.
  Space and astronomy : notable research and discoveries / Kyle Kirkland.
    p. cm. — (Frontiers of science)
  Includes bibliographical references and index.
  ISBN 978-0-8160-7445-7
  1. Space astronomy. 2. Outer space—Exploration. 3. Discoveries in science. I. Title.

  QB136.K57 2010
  520—dc22   2009032803

Facts On File books are available at special discounts when purchased in bulk quantities for businesses, associations, institutions, or sales promotions. Please call our Special Sales Department in New York at (212) 967-8800 or (800) 322-8755.

You can find Facts On File on the World Wide Web at http://www.factsonfile.com

Text design by Kerry Casey
Composition by Mary Susan Ryan-Flynn
Illustrations by Melissa Ericksen
Photo research by Tobi Zausner, Ph.D.
Cover printed by Bang Printing, Inc., Brainerd, Minn.
Book printed and bound by Bang Printing, Inc., Brainerd, Minn.
Date printed: June 2010
Printed in the United States of America

10 9 8 7 6 5 4 3 2 1

This book is printed on acid-free paper.

CONTENTS

PREFACE

Discovering what lies behind a hill or beyond a neighborhood can be as simple as taking a short walk. But curiosity and the urge to make new discoveries usually require people to undertake journeys much more adventuresome than a short walk, and scientists often study realms far removed from everyday observation—sometimes even beyond the present means of travel or vision. Polish astronomer Nicolaus Copernicus's (1473–1543) heliocentric (Sun-centered) model of the solar system, published in 1543, ushered in the modern age of astronomy more than 400 years before the first rocket escaped Earth's gravity. Scientists today probe the tiny domain of atoms, pilot submersibles into marine trenches far beneath the waves, and analyze processes occurring deep within stars.

Many of the newest areas of scientific research involve objects or places that are not easily accessible, if at all. These objects may be trillions of miles away, such as the newly discovered planetary systems, or they may be as close as inside a person's head; the brain, a delicate organ encased and protected by the skull, has frustrated many of the best efforts of biologists until recently. The subject of interest may not be at a vast distance or concealed by a protective covering, but instead it may be removed in terms of time. For example, people need to learn about the evolution of Earth's weather and climate in order to understand the changes taking place today, yet no one can revisit the past.

Frontiers of Science is an eight-volume set that explores topics at the forefront of research in the following sciences:

- biological sciences
- chemistry

- computer science
- Earth science
- marine science
- physics
- space and astronomy
- weather and climate

The set focuses on the methods and imagination of people who are pushing the boundaries of science by investigating subjects that are not readily observable or are otherwise cloaked in mystery. Each volume includes six topics, one per chapter, and each chapter has the same format and structure. The chapter provides a chronology of the topic and establishes its scientific and social relevance, discusses the critical questions and the research techniques designed to answer these questions, describes what scientists have learned and may learn in the future, highlights the technological applications of this knowledge, and makes recommendations for further reading. The topics cover a broad spectrum of the science, from issues that are making headlines to ones that are not as yet well known. Each chapter can be read independently; some overlap among chapters of the same volume is unavoidable, so a small amount of repetition is necessary for each chapter to stand alone. But the repetition is minimal, and cross-references are used as appropriate.

Scientific inquiry demands a number of skills. The National Committee on Science Education Standards and Assessment and the National Research Council, in addition to other organizations such as the National Science Teachers Association, have stressed the training and development of these skills. Science students must learn how to raise important questions, design the tools or experiments necessary to answer these questions, apply models in explaining the results and revise the model as needed, be alert to alternative explanations, and construct and analyze arguments for and against competing models.

Progress in science often involves deciding which competing theory, model, or viewpoint provides the best explanation. For example, a major issue in biology for many decades was determining if the brain functions as a whole (the holistic model) or if parts of the brain carry out specialized functions (functional localization). Recent developments in brain imaging resolved part of this issue in favor of functional localization by showing that specific regions of the brain are more active during certain tasks. At the same time, however, these experiments have raised other questions that future research must answer.

The logic and precision of science are elegant, but applying scientific skills can be daunting at first. The goals of the Frontiers of Science set are to explain how scientists tackle difficult research issues and to describe recent advances made in these fields. Understanding the science behind the advances is critical because sometimes new knowledge and theories seem unbelievable until the underlying methods become clear. Consider the following examples. Some scientists have claimed that the last few years are the warmest in the past 500 or even 1,000 years, but reliable temperature records date only from about 1850. Geologists talk of volcano hot spots and plumes of abnormally hot rock rising through deep channels, although no one has drilled more than a few miles below the surface. Teams of neuroscientists—scientists who study the brain—display images of the activity of the brain as a person dreams, yet the subject's skull has not been breached. Scientists often debate the validity of new experiments and theories, and a proper evaluation requires an understanding of the reasoning and technology that support or refute the arguments.

Curiosity about how scientists came to know what they do—and why they are convinced that their beliefs are true—has always motivated me to study not just the facts and theories but also the reasons why these are true (or at least believed). I could never accept unsupported statements or confine my attention to one scientific discipline. When I was young, I learned many things from my father, a physicist who specialized in engineering mechanics, and my mother, a mathematician and computer systems analyst. And from an archaeologist who lived down the street, I learned one of the reasons why people believe Earth has evolved and changed—he took me to a field where we found marine fossils such as shark's teeth, which backed his claim that this area had once been under water! After studying electronics while I was in the air force, I attended college, switching my major a number of times until becoming captivated with a subject that was itself a melding of two disciplines—biological psychology. I went on to earn a doctorate in neuroscience, studying under physicists, computer scientists, chemists, anatomists, geneticists, physiologists, and mathematicians. My broad interests and background have served me well as a science writer, giving me the confidence, or perhaps I should say chutzpah, to write a set of books on such a vast array of topics.

Seekers of knowledge satisfy their curiosity about how the world and its organisms work, but the applications of science are not limited to intellectual achievement. The topics in Frontiers of Science affect society on a multitude of levels. Civilization has always faced an uphill battle to procure scarce resources, solve technical problems, and maintain order. In modern times, one of the most important resources is energy, and the physics of fusion potentially offers a nearly boundless supply. Technology makes life easier and solves many of today's problems, and nanotechnology may extend the range of devices into extremely small sizes. Protecting one's personal information in transactions conducted via the Internet is a crucial application of computer science.

But the scope of science today is so vast that no set of eight volumes can hope to cover all of the frontiers. The chapters in Frontiers of Science span a broad range of each science but could not possibly be exhaustive. Selectivity was painful (and editorially enforced) but necessary, and in my opinion, the choices are diverse and reflect current trends. The same is true for the subjects within each chapter—a lot of fascinating research did not get mentioned, not because it is unimportant, but because there was no room to do it justice.

Extending the limits of knowledge relies on basic science skills as well as ingenuity in asking and answering the right questions. The 48 topics discussed in these books are not straightforward laboratory exercises but complex, gritty research problems at the frontiers of science. Exploring uncharted territory presents exceptional challenges but also offers equally impressive rewards, whether the motivation is to solve a practical problem or to gain a better understanding of human nature. If this set encourages some of its readers to plunge into a scientific frontier and conquer a few of its unknowns, the books will be worth all the effort required to produce them.

ACKNOWLEDGMENTS

Thanks go to Frank K. Darmstadt, executive editor at Facts On File, and the rest of the Facts On File staff for all their hard work, which I admit I sometimes made a little bit harder. Thanks also to Tobi Zausner for researching and locating so many great photographs. I also appreciate the time and effort of a large number of researchers who were kind enough to pass along a research paper or help me track down some information.

INTRODUCTION

Most astronomical objects are so far away that many early scientists and philosophers despaired of ever acquiring direct knowledge of their composition. For example, the French philosopher Auguste Comte (1798–1857), noted for his pioneering work on sociology and the nature of science, argued that knowledge can stem only from empirical methods—humans can know only what can be sensed or experienced. Comte wrote extensively on his philosophy in the 1830s and early 1840s and asserted that there are some things scientists will never learn. Included in this list was the nature and composition of astronomical bodies.

But advances in science quickly proved Comte and other skeptics wrong. The German scientists Gustav Kirchhoff (1824–87) and Robert Bunsen (1811–99) developed the field of *spectroscopy* in the 1850s and '60s. These and other researchers showed how light could be analyzed in terms of its spectrum—its frequency components—to reveal the chemical elements and compounds that had emitted the light or had absorbed certain parts of it. Using faint lines in the spectrum of the Sun and other stars, astronomers discovered the composition of these bodies and proved that stars were remote bodies with properties similar to the Sun. In 1868, the British astronomer Sir Joseph Norman Lockyer (1836–1920) and the French astronomer Pierre-Jules Janssen (1824–1907) discovered helium in the Sun before this element was found on Earth!

Space and Astronomy, one volume of the Frontiers of Science set, is devoted to scientists who explore the frontiers of space and astronomy—and often find things no one has ever seen before on Earth. The term *astronomy* derives from Greek words *astron,* meaning "star," and *nomos,* "law," and is the study of objects in space beyond Earth and the laws that govern their properties and interactions. This book discusses six topics that encompass a wide range of space and astronomy.

Daring explorers like to study distant frontiers by venturing out into them, but other people prefer to study distant frontiers by bringing them, or representative samples, a little closer to the laboratory. Both options are pursued in the fields of space and astronomy. Some of the chapters of this book focus on space exploration and development, while others are more concerned with astronomy. But space exploration and astronomy are intricately linked: Astronomy guides and informs astronauts (and future colonists) along with unmanned probes that explore space, and these missions return valuable data that advance astronomical knowledge.

Each chapter explores one of the frontiers of space exploration or astronomy. The development of technology such as rocket propulsion or orbiting telescopes is critical in both fields, so this aspect of space and astronomy features prominently throughout the book. Reports published in journals, presented at conferences, and issued in news releases describe research problems and technological developments of interest in space and astronomy, as well as how scientists and engineers are studying them. This book discusses a selection of these reports—unfortunately there is room for only a fraction of them—which offer the student and other readers insight into the methods and applications of astronomy and space exploration.

Astronomy and space exploration can be complicated subjects. Students need to keep up with the latest developments in these quickly moving fields, but they have difficulty finding a source that explains the basic concepts while discussing the background and context essential for the "big picture." *Space and Astronomy* describes the evolution of each of the six main topics it covers and explains the problems that researchers are currently investigating as well as the methods they are developing to solve them.

Chapter 1 describes the search for planets beyond the solar system and the types of planetary systems astronomers have already found. For many centuries, astronomers knew of only one planetary system in the universe, containing Earth and seven other planets (the International Astronomical Union demoted Pluto in 2006, labeling the formerly classified planet as a "dwarf"). But the galaxy is home to billions of stars around which planets may circle. Astronomers have taken advantage of recent advances in instrumentation to detect hundreds of planets that orbit distant stars. None of these planets strikingly resemble Earth, but ongoing projects are continuing the hunt.

Finding distant Earth-like planets would be a milestone in astronomy, but their existence is not essential in the effort to colonize space. As discussed in chapter 2, plenty of opportunities present themselves within the solar system. The *International Space Station* is a large satellite orbiting Earth that has been continuously inhabited by different crews since 2000. Space exploration generates a lot of scientific research opportunities as well as the excitement of adventure, but incentives for extended journeys, and perhaps eventually colonization, include economic benefits and the relief of environmental stress on Earth. These benefits may become necessities if the human population continues to grow at a rapid pace, and the *International Space Station* may be only the first step in a permanent and expanding human presence in space. This chapter discusses potential habitats in orbit or on other worlds such as the Moon and Mars, in addition to the benefits and hazards associated with each.

Some people yearn to go even farther into space. The stars are vastly distant—the closest is the triple star system Alpha Centauri at a distance of 25.6 trillion miles (41.2 trillion km), which is more than 63,000 times the distance from Earth to the Sun. Such a trip would require many times the human lifespan if one's ride consisted of rocket engines that propel vehicles such as the space shuttle and various missiles. But research on new schemes of propulsion, along with unfamiliar consequences encountered at high speeds, such as the slowing of time, may one day permit journeys to the stars. Chapter 3 describes propulsion science and technology and explains the relevant theories of physics.

Physics theory also plays an important role in the subject of chapter 4. The German-American physicist Albert Einstein (1879–1955) developed a theory in 1916 that is the most accurate depiction to date of the force of gravitation. This theory predicts the existence of gravitational waves rippling through space and time, similar to electromagnetic waves such as light. Massive objects or systems undergoing abrupt changes theoretically emit these waves. Although scientists have found indirect evidence for gravitational waves, researchers have yet to make a direct measurement because the waves have such a tiny magnitude. The effort to design and build instruments sensitive enough to detect gravitational waves is an important frontier of science because researchers can use these instruments to study and test gravitation theories. Gravitational wave "telescopes" would also be a boon to astronomers who could use them to study the nature and motion of distant objects.

One of the most important questions about the universe involves the formation and evolution of galaxies, the topic of chapter 5. Many galaxies have one of two different structures—a globular shape or a disk with spirals—and the rest lack any sort of regular structure. Lurking within the center of many galaxies is a strange object called a *black hole* with such a strong gravitational field that nothing can escape it. Researchers are piecing together the history of the universe in order to understand how galaxies came to be, their relationship with black holes, and their eventual fate.

Astronomers have recently discovered an even more puzzling aspect of the universe—most of it is invisible. Certain gravitational effects in galaxies indicate the existence of a lot more mass than can be seen. Another set of observations suggest the existence of a strange kind of energy that is pushing the universe apart. Although alternative explanations may emerge, scientists are currently focusing on the search for the missing mass and an unusual form of energy, as discussed in chapter 6. No one has a firm idea what sort of substances compose this hidden portion of the universe, but their nature and eventual discovery—or proof of an alternative hypothesis—is an extremely active research topic.

The hidden portion of the universe is a typical frontier—obscure, distant, and enticing to those who love a challenge. All of the chapters of this book guide the reader to a special place beyond the grasp of the present state of science. As scientists struggle to extend their reach, some of these mysteries may soon be solved, but some may remain unsolved for a long time, awaiting the ideas and breakthroughs of a future generation of researchers.

Extrasolar Planets: Worlds beyond the Solar System

In August 2006, the solar system lost one of its nine planets—not because of a cataclysmic event, but because astronomers at the 2006 International Astronomical Union (IAU) General Assembly decided upon a new definition for the term *planet*. IAU astronomers agreed that a planet is a body with sufficient mass and gravity to be round or nearly round and whose orbit around the Sun is a clear path (no other significant bodies should share a planet's orbit). Under this definition, Pluto, a small body whose highly eccentric orbit takes it into a zone called the Kuiper Belt occupied by many other orbiting bodies, fails to meet the definition. The solar system now officially has eight planets. Pluto, which astronomers consider to be a "dwarf planet," is just one of the many Kuiper Belt bodies.

Defining a term is important in using it in a scientifically precise manner. There are plenty of bodies in the solar system, and as of 2006 astronomers have started classifying them based on distinguishing characteristics rather than on tradition. This is also important in a more general study of planetary systems since astronomers have recently begun discovering small objects orbiting other stars in the galaxy. Although the IAU definition applies only to the solar system, the objects orbiting other stars are in the same size and mass range as the heavier planets in the solar system, so scientists call these objects planets. As of July 2009, the PlanetQuest Web site lists 353 *extrasolar planets,* or *exoplanets,* orbiting 297 stars (some of which have more than one planet).

None of these planets is much like Earth. The majority are a lot like Jupiter, which is a planet that is much larger than Earth and composed mostly of gases. But the absence of Earth-like planets is probably due to the limits of the methods of detection. Stars other than the Sun lie at extremely vast distances; the closest star (besides the Sun) is Alpha Centauri, a triple-star system that is 25.6 trillion miles (41.2 trillion km) away, which is more than 63,000 times the distance from Earth to the Sun. The universe is so big that astronomers describe distances in a unit called a *light-year,* which is the distance light travels (in a vacuum) in one year. One light-year is 5.88 trillion miles (9.46 trillion km), so Alpha Centauri resides about 4.35 light-years away. Most other stars are a lot farther away than this.

Since stars shine brightly but the much smaller planets are visible only by the tiny amount of light they reflect, spotting extrasolar planets is quite difficult. People have likened the task to that of seeing a firefly next to a lighthouse or locating a lighted match in a forest fire. Current techniques can detect only large extrasolar planets, usually orbiting close to the star. Most of the time, astronomers cannot directly see an extrasolar planet but instead discover it from the effects it has on the star or the surrounding dust or gas.

But these extrasolar planetary systems can tell astronomers a lot about planets and how they form—it is always desirable to have more than one example to study. The existence of other planets leads to interesting questions about life elsewhere in the universe. People have long wondered if Earth is the only harbor of life and if human beings have fashioned the galaxy's or perhaps even the universe's only civilization. The study of extrasolar planets will go a long way in helping to answer these questions. This chapter describes how astronomers detect these remote bodies, what has been learned about them thus far, and the techniques scientists are developing to improve their equipment in the attempt to find an extrasolar "Earth."

INTRODUCTION

Ancient peoples used the stars and their motion to construct calendars and to help them navigate the seas. The positions of most of these bright points of light are fixed relative to one another, forming constellations that move slowly across the sky, as if attached to a celestial sphere. But some of the "stars" march to a different beat, wandering among the fixed stars. These points of light became known as planets, after a Greek word that means "wanderer."

The idea of worlds other than Earth had to wait until the general acceptance that there was nothing unique about this planet and the Sun around which Earth and the other planets orbit. Uncritically examined beliefs dominated early notions of astronomy and cosmology, which placed Earth at the center of the universe. The Polish scientist Nicolaus Copernicus (1473–1543) proposed a heliocentric theory of the solar system, in which Earth and the other planets revolved around the Sun. This theory paved the way for astronomical advances, including the realization that planets might exist outside of the solar system.

But Copernican astronomy became mired in controversy at the time the theory was published in 1543. One of the most zealous of the early advocates of the heliocentric theory, the Italian philosopher Giordano Bruno (1548–1600), incurred the ire of religious authorities, who executed him. Among Bruno's controversial ideas was the belief that the universe was infinite and populated with numerous worlds.

In 1610, the Italian scientist Galileo Galilei (1564–1642) aimed a newly fashioned telescope at Jupiter and found tiny points of light that seemed to be orbiting it—some of the moons of Jupiter. Galileo—generally known by his first name—also staunchly supported the heliocentric theory, although he publicly recanted this belief in 1633 when powerful church authorities, who held different views, threatened him. But astronomers gradually confirmed Copernicus's ideas and realized that the stars that dot the night sky are faraway Suns. The other planets of the solar system, although appearing to the unaided eye as points of light, can be observed as spherical worlds with the use of a powerful telescope.

Interest in other worlds sparked a class of literature that came to be known as science fiction. In 1898, the British author Herbert George (H. G.) Wells published *The War of the Worlds,* in which creatures from Mars invade Earth. Wells and other writers popularized the notion of life beyond Earth. Many of the stories involved creatures from planets in the "local" solar system, but the imagination of other writers extended to include creatures and civilizations from distant parts of the galaxy. Such stories fostered a belief in the existence of extrasolar planets

The eight planets of the solar system, drawn to scale, and a portion of the Sun, also drawn to scale *(Detlev van Ravenswaay/Photo Researchers, Inc.)*

since a planet seems best suited to harbor life that resembles life on Earth.

But telescopic studies and space probes sent to other planets in the solar system diminished prospects of finding life on these worlds and virtually eliminated the belief that advanced civilizations existed on them. In some sense, early ideas about Earth are correct—the planet does occupy a special place, at least in the solar system, since its orbit and composition permit rich, watery environments. None of the other planets in the solar system enjoy temperatures at which water can exist on the surface of the planet. While the solar system contains four rocky, terrestrial planets—Mercury, Mars, Earth, and Mars—only Earth is comfortable for life as it is currently understood.

The Milky Way galaxy houses the solar system and an estimated 300 billion other star systems. (And there are many more galaxies in universe.) Many of these star systems are binary, meaning that they contain two stars orbiting each other, and some systems contain three stars. No one knows how many of these stars have planets. This is one of the more significant issues that this frontier of science will address. Most astronomers believe that planets are a natural accompaniment to star formation if sufficient material is available, and the study of extrasolar planets will help answer this question, as described in the section below, titled "Formation of Planets." Even if only a small fraction of these stars have planets, there are a huge number of planets in the Milky Way galaxy alone.

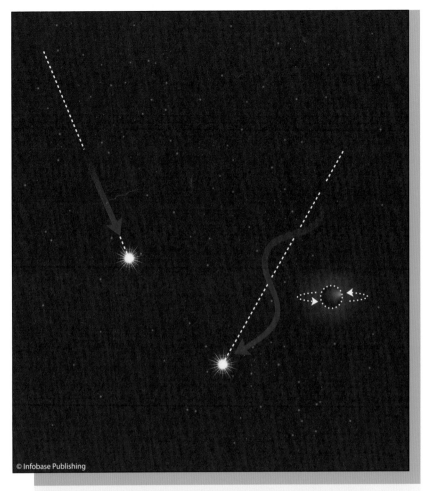

A star as well as its planets orbit the system's center of mass, which introduces a slight wobble in the star's motion, as depicted in the right half of the figure.

Astronomers have discovered the vast majority of extrasolar planets indirectly by the effects they have on the stars they orbit. Consider gravity, for example. The Sun is much more massive than its planets—the Sun's mass is about 750 times that of the planets combined—so people speak of planets orbiting or circling the Sun. But gravitational attraction is mutual, and in reality, a planet and its star move around a common point, which is the center of their mass. In the case of Earth and the Sun, the center of mass is extremely close to the center of the Sun, which would produce no detectable wobble. But the center of mass is slightly farther away from the Sun's center for the more massive planets. The planets' gravitational tug on the

Arecibo Observatory

Located a short distance from Arecibo, a town in Puerto Rico, the radio telescope at the Arecibo Observatory is the largest single telescope in the world. Nestled in a depression in the ground caused by a sinkhole, the dish is 1,000 feet (305 m) in diameter. The telescope's surface consists of nearly 40,000 perforated aluminum panels, each of which is about 3 feet (0.9 m) by 6 feet (1.8 m), held up by a network of steel cables stretching across the sinkhole.

William E. Gordon, a professor at Cornell University in New York, was one of the main organizers for the development of the observatory, which formally opened on November 1, 1963. Today, the observatory is part of the National Astronomy and Ionosphere Center, operated by Cornell University. About 140 personnel work at the observatory, and each year 200 scientists visit the facility and use the telescope for various research projects, as did Alex Wolszczan and his colleagues in the early 1990s when they found evidence for planets circling a pulsar.

Similar to an optical reflecting telescope, the dish of a radio telescope reflects and focuses radio waves, which are electromagnetic waves belonging to the low-frequency, long-wavelength portion of the spectrum. A radio receiver turns the waves into a signal that astronomers study to learn more about the radio wave source. The larger the dish, the more radiation it can collect, which means larger instruments can detect weaker signals. Since radio waves are long waves— about 100,000 times longer than visible light—they are not easily scattered by imperfections in the dish, allowing radio telescopes to be huge without being so expensive. (A flaw in

Sun introduces a slight wobble in its motion in the galaxy, as illustrated in the figure on page 5, since the Sun orbits the solar system's center of mass.

In the 1960s, the Dutch-American astronomer Peter van de Kamp (1901–95) studied the motion of Barnard's star, a small star whose

The radio telescope at the Arecibo Observatory *(NAIC-Arecibo Observatory, a facility of the NSF)*

the *Hubble Space Telescope* that was only a 50th of the diameter of a human hair needed special corrective optics in 1993.) But for the same reason, radio telescopes have poorer resolution.

Astronomers have made many discoveries with the Arecibo Observatory. In addition to Wolszczan's finding, scientists at Arecibo established Mercury's rotational speed, discovered the first pulsar in a binary system, mapped the distribution of galaxies in the universe, and much else.

distance of six light-years makes it the second-closest star system to Earth (after the triple star system of Alpha Centauri). (Barnard's star was named for its discoverer, the American astronomer Edward E. Barnard [1857–1923].) Van de Kamp believed that he found a wobble in the motion of Barnard's star, indicating the presence of an unseen companion. His calculations suggested that the companion was a large planet several times more massive than Jupiter. But subsequent observations showed that this wobble was an artifact, due to faulty measurements rather than to a planet orbiting the star.

But van de Kamp's persistence and optimism about extrasolar planets was warranted. A few decades later, scientists succeeded in detecting the first planet beyond the solar system.

PULSAR PLANETS

Although earlier studies found hints of extrasolar planets that were subsequently confirmed, the first quickly confirmed extrasolar planet was discovered in 1991. Alex Wolszczan, an astronomer at Pennsylvania State University, and his colleagues were using the giant radio telescope at the Arecibo Observatory in Puerto Rico to study a *pulsar*—a rotating *neutron star*. A pulsar is a dense remnant of a supernova. During a star's lifetime, nuclear reactions taking place in the core provide energy for the star to shine and to counteract the force of gravity. When the nuclear fuel runs out, some of the more massive stars die in a spectacular explosion called a supernova, which throws off some of the mass. Gravity collapses the remaining mass into a small space—astronomers believe neutron stars typically have a radius of perhaps 12.4 miles (20 km)—which generates such enormous pressure that protons and electrons get combined into neutrons. Neutron stars have strong magnetic fields and can rapidly rotate, sending out pulses of electromagnetic radiation, which gives pulsars their name (short for pulsating stars). Astronomers observe these pulses as flashes that occur at a certain point in the pulsar's rotation. Pulsars are similar to lighthouses, emitting a beam of radiation—usually radio waves instead of light—which observers can see at regular intervals that are governed by the rate of rotation. These intervals have clockwork precision.

With the sensitive Arecibo radio telescope, Wolszczan timed the pulses of a pulsar in the Virgo constellation. The presence of bodies orbiting the pulsar will make it wobble around the system's center of mass. Although most of the mass is in the pulsar, the result is a scarcely noticeable wobble in the pulsar's motion, and this will affect the flashes as seen on Earth—the regularity of the pulses is slightly perturbed. Wolszczan found millisecond departures from the pulsar's "clock," indicating the existence of orbiting objects. He announced the finding in 1992.

Arecibo Observatory's radio telescope was an essential instrument for this research because of its great size and sensitivity. The sidebar on page 6 provides more information on this observatory.

Wolszczan used pulsar timing to determine that three bodies were orbiting the pulsar. The mass of two of the bodies slightly exceeded Earth's mass, while the other body was much lighter. All three orbits were closer to the pulsar than Mercury is to the Sun.

The bodies around this pulsar are not necessarily planets if astronomers were to apply IAU's 2006 definition to extrasolar systems, though most people refer to them as such. (There is as yet no fixed definition of the term *extrasolar planet*.) But one thing is for certain—these bodies orbit a dead star, and no life could exist on them, at least not life similar to Earth organisms. Two of the planets are Earth-like in size, but the resemblance ends there. Perhaps these worlds may have been somebody's home before the cataclysmic death of this star, but not now.

USING RADIAL VELOCITY TO DETECT PLANETS

Although the earliest confirmed worlds beyond the solar system orbit a dead star, astronomers quickly found planets elsewhere. In 1995, the Swiss astronomers Michel Mayor and Didier Queloz at the Geneva Observatory discovered planets orbiting a star on the *main sequence*. The main sequence is a band that appears when stars are plotted in a diagram based on spectrum and luminosity. These diagrams are known as Hertzsprung-Russell (H-R) diagrams, named after the Danish astronomer Ejnar Hertzsprung (1873–1967) and the American astronomer Henry Russell (1877–1957). Stars on the main sequence are "alive," shining with the energy created in nuclear fusion reactions in their hot, dense centers.

Mayor and Queloz studied 51 Pegasi, a star about 50 light-years from Earth in the constellation Pegasus (the winged horse). This star is similar to the Sun, having roughly the same mass and comparable spectral

Doppler Effect

Suppose a stationary source emits a steady wave such as light or sound, as shown in part (A) of the figure. The frequency of the wave is its number of cycles per second. For example, the frequency of green light is roughly 540 trillion *hertz* (cycles per second). Wavelength is the distance between crests (or any two corresponding points on the wave's cycle), and it is inversely related to the wave's frequency—high-frequency waves have short wavelengths, while lower frequencies have longer ones.

Now suppose the source is moving, as shown in part (B) of the figure. To an observer standing in front (to which the source is moving), the emitted waves are closer together. The reason for this is as follows: As the source is moving closer to the observer, the source has covered some distance between the time it emits one peak of the wave and the time it emits the next peak, which means that the distance between these points—which is equal to the wavelength—is shorter as viewed by a stationary observer. The result is that the wavelength decreases and the frequency increases. An opposite shift occurs for an observer from which the object is receding—the wavelength increases and the frequency decreases.

An audible example of the Doppler effect can be heard when an approaching or receding train blows its whistle. Suppose an observer is standing on a station platform as a train moves past. If the train blows its whistle while approaching the platform, the frequency, or pitch, of the sound

characteristics. The researchers found periodic variations in the star's *radial velocity* that suggested the presence of a planet in this system.

Radial velocity of an object is its relative velocity along the observer's line of sight, or in other words, how fast the object is moving toward or away from the observer. When the object is emitting a wave such as

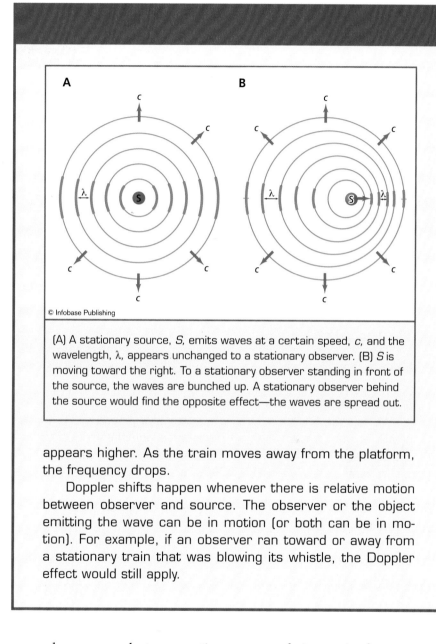

© Infobase Publishing

(A) A stationary source, S, emits waves at a certain speed, c, and the wavelength, λ, appears unchanged to a stationary observer. (B) S is moving toward the right. To a stationary observer standing in front of the source, the waves are bunched up. A stationary observer behind the source would find the opposite effect—the waves are spread out.

appears higher. As the train moves away from the platform, the frequency drops.

Doppler shifts happen whenever there is relative motion between observer and source. The observer or the object emitting the wave can be in motion (or both can be in motion). For example, if an observer ran toward or away from a stationary train that was blowing its whistle, the Doppler effect would still apply.

a sound wave or an electromagnetic wave, any relative motion between object and observer will shift the frequency of the wave. As described in the sidebar above, this effect is known as the *Doppler effect* or the Doppler shift, named after the Austrian physicist and mathematician Christian Doppler (1803–53), who first explained it.

Suppose a star has a companion such as a planet, or perhaps more than one planet. Although stars are generally much heavier, both objects will orbit about the system's center of mass. To an observer near the orbital plane—the geometrical plane, or two-dimensional area, containing the planet's orbit—the star will appear to move back and forth as it moves around the center of mass, as shown in the figure. The result

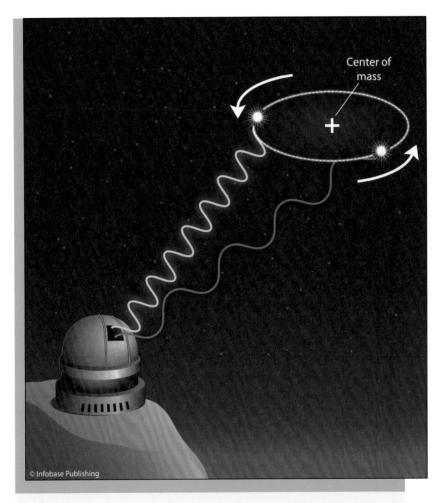

Due to the Doppler effect, the star's radiation has a blueshift—higher frequency and lower wavelength—when the star's orbital motion brings it closer to the observatory, as shown in the upper left portion of the figure. A redshift occurs when the star is receding.

is that the frequency of the light appears shifted due to the Doppler effect. When the star is approaching the observer, its light appears bluish, or shifted to the high-frequency end of the spectrum; a receding star's light is redshifted.

To an observer standing directly above, looking down on the star, or directly below, looking up, the star would be moving in circles. This would not change the radial velocity because the star would not be moving toward or away from the observer. But to all other observers, the gravitational attraction would change the star's radial velocity.

Radial velocity is an indirect but powerful method of finding planetary companions of a star. Mayor and Queloz discovered periodic variations in the velocity of 51 Pegasi, as revealed in its spectrum, which indicated the presence of another mass in the system. The magnitude of these variations depends on the gravitational influence of the other body, which is a function of the mass of the companion and the distance at which it orbits. These variations and the interval over which they occur give researchers a way to compute the companion's orbit. Mass can also be calculated, but not precisely. In addition to possible measurement error, researchers do not know at what angle they are viewing the orbital plane. Because they are unsure of the angle, astronomers cannot exactly determine the properties of the orbits, so there is some estimation involved.

The *orbital period*—the time required for one complete revolution—of the companion of 51 Pegasi is an astonishingly short 100 hours—the "year" of this planet corresponds to about 4.2 days on Earth. The planet is located about 5,000,000 miles (7,900,000 km) from the star, slightly more than a twentieth of the distance from Earth to the Sun (which is called the *astronomical unit,* abbreviated au). Astronomers estimate the mass to be about half that of Jupiter. The name of the planet is 51 Pegasi b. This name follows a pattern of calling planets by the name of their star and an appended letter, beginning with *b* for the first planet discovered around that star, *c* for the next, and so on. (The letter *a* is reserved for the star itself.) Geoffrey Marcy and Paul Butler used the Lick Observatory to confirm the finding of Mayor and Queloz shortly after the discovery.

Astronomers did not discover extrasolar planets sooner because radial velocity measurements are difficult to make and subject to many errors. Only recently could astronomers make reliable measurements.

While radial velocity is not the only method of discovering extrasolar planets, the majority of the planets have been found with this technique. Other techniques include astrometry—measuring a star's wobble, as described above, and pulsar timing to detect planets around pulsars. When a planet's orbit crosses in front of the star, astronomers may detect a periodic change in brightness, a technique known as the transit method. Gravitational microlensing is a technique based on the gravitational field of a star acting as a magnifying lens on the light of a distant star, and the circumstellar disk method uses the spectra of dust surrounding a star to reveal the presence of planets.

FORMATION OF PLANETS

Many of the extrasolar planets discovered thus far are unusual, as is 51 Pegasi b. But this does not necessarily imply that many planetary systems are unlike the solar system in which Earth resides, since the sample of discovered extrasolar planets is skewed due to the detection methods. Most of the planets are gas giants, similar to Saturn and Jupiter. Some, such as 51 Pegasi b, are called *hot Jupiters* because their orbits place them close to the star, and proximity to the star's radiation undoubtedly creates a warm planet. Note that hot Jupiters are easy to detect with the radial velocity method because the variations they cause in the star are quite large. Other categories include icy giants, small planets in orbits near their star, and planets found around pulsars. Astronomers must take care not to confuse extrasolar planets with objects known as *brown dwarfs,* which are much larger than Jupiter but do not have the size to initiate fusion reactions. Brown dwarfs can be more massive than Jupiter by a factor of about 12 times or more, although if the mass is more than about 80 times Jupiter's, the object may begin to shine—and become a star.

Another important distinction between stars and planets is the process of formation. According to currently accepted ideas, stars are born when interstellar material—clouds of dust and gas in the galaxy—begins to condense, perhaps because of some external field or fluctuation. Due to gravitational attraction and angular momentum, the material forms a disk, and particles begin to collect in the center, which becomes hotter. The mass, temperature, and pressure rises, eventually initiating nuclear reactions. A star is born. Near the end of this process, small bodies called protoplanets have formed by gravitational contraction and accretion (the collision and sticking of smaller pieces of matter). These objects cool and become planets, settling into stable orbits and dispersing smaller objects with their gravitational fields. (Jupiter, for instance, exerts considerable influence in and around its orbit.)

But the details of star and planet formation are not yet clear. Planetary systems beyond Earth are starting to provide additional data that astronomers can analyze. Although the many hot Jupiters that have been found may not reflect their abundance in the galaxy, their existence is causing some astronomers to reconsider theories of planet formation. Prevailing concepts suggest that cooler regions of the disk, at least several astronomical units from the star, are most favorable for the formation of gas giants. To explain hot Jupiters, researchers are highlighting the importance of complex interactions that may cause gas giants to migrate inward, decreasing their orbital radius.

In the solar system, rocky planets form the inner portion and gas giants are farther out, in contrast to systems with hot Jupiters. Edward W. Thommes, a researcher at the University of Guelph in Ontario, Canada, and his colleagues have conducted simulations on a computer to study the process of planet formation. They found that when the orbits are elongated—they have a high *eccentricity*—gas giants tend to migrate toward the center. Many of the extrasolar planets' orbits are elongated, supporting this result. By adjusting the size and nature of the disk, along with other parameters, the researchers ran numerous simulations. The results suggested that the hot-Jupiter kind of planetary system is more common than the local solar system. As Thommes and his colleagues wrote in their paper published in *Science* in 2008, "Analogs to our own solar system do not appear to be common."

EXTRASOLAR EARTHS

The probability that a planetary system will develop a structure similar to the solar system may not be all that high, but with billions of opportunities throughout the galaxy, more than a few should arise. Assuming that the solar system is not unique, then it seems likely that planets with characteristics similar to Earth should exist. But the only way to prove this is to find one. Although no one has yet found an extrasolar Earth, the hunt is on.

National Aeronautics and Space Administration (NASA)

NASA got its start on October 1, 1958. However, the primary motivation for the organization's establishment occurred almost exactly a year earlier. On October 4, 1957, the former Union of Soviet Socialist Republics (USSR) launched the first artificial satellite, *Sputnik.* (The term *sputnik* is Russian for traveling companion.) The satellite did little but orbit Earth and emit beeps, but the technological achievement and its potential uses staggered the world. At the time, the United States and the USSR were engaged in a cold war of differing political ideologies as well as economic, social, and technological competition. USSR's success in space instilled fear in many Americans that the United States was falling behind in science and technology. Recognizing the impact of science in World War II, especially the development of fearsome nuclear weapons, the United States rushed to catch up. The creation of NASA came about by upgrading the National Advisory Committee for Aeronautics (NACA), NASA's predecessor.

One of NASA's many achievements was the Apollo project, in which astronauts Neil Armstrong and Edwin "Buzz" Aldrin were the first men to walk on the Moon on July 20, 1969. NASA also operates and maintains the fleet of space shuttles, which carry astronauts and payloads such as satellites into orbit. The advancement of science is also a major goal of the agency. In 1990, for example, the space shuttle *Discovery* carried the *Hubble Space Telescope* into orbit. This telescope, for which NASA partnered with the European Space Agency (ESA), operates above the interference of the atmosphere, giving astronomers unprecedented views.

Alan L. Bean, NASA astronaut of the *Apollo 12* mission, steps down the ladder of the lunar module on his way to join astronaut Charles Conrad, Jr., on the Moon. *(NASA/Johnson Space Center)*

Prospects for future missions are equally bold and pioneering. NASA is currently preparing or developing plans to return to the Moon, perhaps establishing a permanent station there, as well as sending a manned mission to Mars. Many new science projects are also in the process of launching or in preparation. Some of these projects will be featured in the chapters of this book.

Finding an Earth-like planet would be stimulating for several reasons. Such a planet is the most likely place to harbor *extraterrestrial life*—life beyond Earth. Biologists do not understand how life arose on Earth, but fossil evidence suggests it did so quickly after the planet's birth. If this is the case, then the rapid development of life on Earth encourages the belief that the process is a natural one rather than a wildly improbable event. Given the right conditions—and the right planet—life may be likely or even almost a certainty. The discovery of extraterrestrial life would have a profound effect on science and philosophy.

Even more profound would be the discovery of a technological civilization similar to or more advanced than humans have achieved. The development of modern human society and technology required billions of years from the time life first arose on Earth; this suggests that while life may be common, technological civilizations may be much less probable. Perhaps the development of civilization requires a tortuous and unlikely series of events. If so, human beings might have no counterpart elsewhere in the universe.

An absence of evidence for extraterrestrial civilizations is also suggestive. The universe is about 14 billion years old (see chapter 4), and the Milky Way galaxy is not much younger—some of the oldest stars in the galaxy were born about 13 billion years ago. In all that time, one might expect plenty of civilizations to arise. Yet if the galaxy teems with civilizations, they are well hidden and have left no unambiguous signs anywhere. Perhaps technological civilizations rarely arise, and perhaps they do not survive for very long when they do.

Determining whether humans are alone in the universe would most likely have an important effect on how people think about themselves and their society. If other civilizations exist, humans will have opportunities to exchange knowledge and compare notes; if not, the realization that human civilization is the only game in town will evoke a variety of emotions, perhaps including caution and pride.

One of the most important steps in deciding this issue is to search for extrasolar Earths. The *National Aeronautics and Space Administration* (NASA), a U.S. agency devoted to space exploration and related sciences and technologies, is investing a great deal of time and energy into this search. NASA boasts many accomplishments in the past and wields considerable influence and resources. The sidebar on page 16 provides more information about this large agency.

NASA's *Hubble Space Telescope* has made numerous contributions to extrasolar planet research. The space telescope cannot generally view extrasolar planets directly, but it can sometimes provide tantalizing glimpses. On November 27, 2001, for instance, the Space Telescope Science Institute issued a news release announcing that astronomers had made direct measurements of an extrasolar planet's atmosphere. The star of this system, called HD 209458, is 150 light-years distant and resembles the Sun, though the planet, which had been discovered in 1999, is a hot Jupiter. Its orbit is only 3.5 days, at a distance of only 4 million miles (6.4 million km) from the star. During its orbit, the planet passes in front of the star, and in this brief period, light from the star passes through the planet's atmosphere before reaching Earth. Because elements and compounds emit and absorb radiation at specific frequencies, their presence creates characteristic lines that can be seen when observers view the light's spectrum. Although the planet is almost certainly too hot to sustain life, and the astronomers did not analyze the spectrum for biological compounds, they did find the element sodium.

In 2008, astronomers studied the atmosphere of another planet with the *Hubble Space Telescope* and the *Spitzer Space Telescope,* which is tuned to radiation chiefly in the infrared portion of the spectrum. NASA issued a news release on December 9, 2008, announcing that astronomers had detected carbon dioxide on a hot Jupiter, HD 189733b, located 63 light-years away. Scientists studied this planet's atmosphere during the planet's transits, but in this case they studied light from the star alone when the planet was blocked from view (as it traveled behind the star) and compared it to the spectrum that included both star and planet, which allowed them to separate the influences of the planet. Astronomers had earlier found evidence of water vapor and methane in the planet's atmosphere.

KEPLER PROJECT

On March 6, 2009, NASA launched *Kepler.* This satellite gives astronomers an excellent tool to search for extrasolar Earths.

According to the "Overview of the *Kepler Mission,*" posted on NASA's *Kepler* Web site, the mission "is specifically designed to survey our region of the Milky Way galaxy to discover hundreds of Earth-size and smaller planets in or near the habitable zone and determine how many of the billions of stars in our galaxy have such planets." The satellite, named in honor of the German mathematician and astronomer Johannes Kepler (1571–1630), uses a special 3.1-foot (0.95-m) diameter telescope to scan a wide field of view, which contains a huge number of stars. *Kepler* keeps track of the brightness of more than 100,000 stars in this region of the galaxy.

Astronomers are counting on the sensitive measurements of brightness to reveal transits of planets, including small and difficult-to-find ones, such as a rocky planet located an astronomical unit from its star. Transits occur when planets cross in front of the star around which they orbit. When transits take place, the planet blocks a small part of the star's light as seen from Earth, similar to a lunar or solar eclipse. An orbiting planet will periodically dim a star's brightness at an interval related to its orbital period. Larger planets block more light than smaller ones and are therefore easier to detect. An Earth-sized planet may cause changes of a factor of 1/10,000th that last a few hours. In order to be sure of the results, astronomers look for repeated transits, thereby excluding unrelated brightness fluctuations.

When astronomers sifting through *Kepler*'s data find likely candidates, they can compute the planet's orbit from the period—the planet's year, or how long it requires to circle the star—and the star's mass. The size of the star and the amount by which the brightness drops during the planet's transit reveals the size of the planet. From these and other properties of the star system, astronomers can deduce features of the planet, such as its surface temperature.

"Overview of the *Kepler Mission*" notes that, "For a planet to transit, as seen from our solar system, the orbit must be lined up edgewise to us. The probability for an orbit to be properly aligned is equal to the diameter of the star divided by the diameter of the orbit. This is 0.5% for a planet in an Earth-like orbit about a solar-like star." If Earth-like planets are rare, astronomers can expect to examine many stars before finding one. Even if every star has an Earth-like planet, only about one in 200 (0.5 percent) would be likely to be detected with this technique. This is why *Kepler* monitors 100,000 stars. But if pessimism proves unwarranted and Earth-like planets are common, the mission should spot hundreds of them.

But the mission will take time. The idea is to find Earth-like planets, which means that the orbital period will be about a year. This means that the time between transits will also be about a year. As "Overview of the *Kepler Mission*" states, "To reliably detect a sequence one needs four transits. Hence, the mission duration needs to be at least three and one half years."

Kepler supplements a previously launched mission called COROT, which stands for *Convection Rotation and Planetary Transits*. Launched on December 27, 2006, and led by the French space agency CNES (Centre National d'Études Spatiales—National Center for Space Research), part of COROT's mission is to monitor star brightness due to planet transits. On February 3, 2009, the European Space Agency (ESA) announced that COROT had found a small planet with an orbital period of 20 hours. According to the news release, "The amazing planet is less than twice the size of Earth and orbits a Sun-like star. Its temperature is so high that it is possibly covered in lava or water vapour."

EXTRATERRESTRIAL BIOLOGY— LIFE BEYOND EARTH

One of the most important factors motivating the search for extrasolar planets resembling Earth is the belief that life may exist elsewhere. But no one knows what form an extraterrestrial organism may take, if any such organisms exist.

Imaginative scientists and science fiction writers have speculated on many different forms of life. For example, Robert L. Forward's 1980 novel *Dragon's Egg* depicted interesting and unfamiliar life forms that had evolved on a neutron star. In 1957, Sir Fred Hoyle published the novel *The Black Cloud,* which involved an immense cloud of gas and dust that exhibited intelligence. Both of these authors were scientists, and their inventive creations that bear no relation to Earth biology were nevertheless rooted in science, despite their speculative nature.

Scientists who ponder what kind of life and biological activities may predominate in the universe have thus far only one sample to study— Earth. No one can draw firm statistical conclusions from a sample size of only one. Yet this sample is all that scientists currently have available, and since terrestrial biology is demonstrable, and other life forms are speculative, many researchers choose to focus on what they have learned about life on Earth.

An artist's rendition of a habitable extrasolar planet *(Detlev van Ravenswaay/Photo Researchers, Inc.)*

Organisms on Earth are diverse, including such different creatures as microbes, ants, sharks, and human beings. But biologists have discovered certain features common to all. Water, for instance, is a general requirement. It is an excellent solvent in which nutrients and biological molecules can exist, and all organisms contain some amount of water. (Viruses are exceptions since they contain only nucleic acids and a protein coat, but viruses are not generally considered to be alive. They only become biologically active when they infect cells.) Organisms also contain some form of replicating instructions contained in a long nucleic acid known as deoxyribonucleic acid (DNA). And most organisms need oxygen, although some species, called anaerobes, do not require this element and can die when it is present. Other features of the environment, such as an atmosphere that shields organisms from high-energy radiation and retains warmth for the planet, also seem to be vital.

The ideal spot for life beyond Earth, if it follows terrestrial biology, is a planet with an atmosphere and plenty of water, at least some of which is in liquid form. This requires a temperature above the freezing point of water, which is 32°F (0°C) when the atmospheric pressure is equal to that on Earth's surface, and below the boiling point, 212°F (100°C) at atmospheric pressure. These values change with atmospheric pressure, but as long as the pressure is not too great or too small, life will be possible.

Finding a planet with oxygen in the atmosphere would be a promising sign. On Earth, oxygen in the atmosphere has been produced over the eons by the metabolic activity of plants. A planet with oxygen does not necessarily contain life—oxygen can be produced in other kinds of reactions—but it would be a good bet.

Another important sign of life are *organic compounds.* The term *organic* refers to carbon-containing substances that are involved in life. Organic compounds include carbohydrates, proteins, nucleic acids such as DNA and ribonucleic acid (RNA), vitamins, fats, and hydrocarbons (compounds of hydrogen and carbon). Carbon is critical because it can form long chains needed to produce large, complex structures.

None of the planets in the solar system other than Earth are known to be suitable for terrestrial-like organisms, but some spots, such as sub-surface regions of Mars and Europa (an icy moon of Jupiter), may harbor life. But there may be plenty of candidates in other star systems.

On March 19, 2008, the Hubble Information Centre announced the discovery of the first organic molecule found on an extrasolar planet. Researchers led by Mark Swain at NASA's Jet Propulsion Laboratory used the *Hubble Space Telescope* to find methane on a hot Jupiter, HD 1899733b (the same planet mentioned earlier). They found the substance by analyzing light passing through the planet's atmosphere, in a way similar to the research described above.

Methane is a hydrocarbon and one of the major components of natural gas. But in this case, no life on this extrasolar planet seems possible since the atmosphere of this hot Jupiter has a temperature of about 1,650°F (900°C). These organic molecules are formed by chemical reactions unrelated to life, at least as far as terrestrial biology is concerned.

CONCLUSION

Organic molecules need not be related to life, as is the case for hot Jupiters, but such compounds found on Earth-like extrasolar planets would be promising. Perhaps with the data from *Kepler* scientists will be able to find a few examples. Since *Kepler* uses transits to make its discoveries, these planets may be candidates for the same sort of research in which atmospheric composition may be analyzed. However, extrasolar Earths have a much longer orbital period than hot Jupiters and are much smaller in size. Studying the atmosphere of these distant worlds may be difficult.

Enhancement of existing instruments and technology will improve the situation. For example, NASA plans on launching another space telescope in 2013. This telescope, the *James Webb Space Telescope* (named for a former NASA administrator), will have a mirror with a diameter of 21.3 feet (6.5 m). The optics of the instrument will be optimized for infrared radiation, permitting researchers to see through clouds of dust and examine developing planetary systems, as well as those that have already formed. When this space telescope is launched, astronomers will have an excellent new tool with which to conduct research, including locating and studying extrasolar planets.

Other future projects are concentrating on direct detection of Earth-like extrasolar planets. Imaging these distant planets with existing ground-based instruments is not currently possible, but scientists are considering methods to do so with advanced equipment. One of the techniques most often discussed is known as *interferometry*.

Interferometry is the process of using the interference of waves to make precise measurements. Waves such as electromagnetic radiation obey the principle of superposition, which says that waves occupying the same space will combine. Two waves of the same frequency and amplitude sum to a combination that has twice the amplitude of the individual waves if they are in phase (if their peaks and troughs occur at the same time). The same two waves will cancel each other if the peak of one wave coincides with a trough of the other. An interferometer can use interference of electromagnetic radiation such as light or radio waves to measure distances with high precision. For instance, by splitting a light beam and allowing each part to travel separate paths until they meet at some common point, an interferometer can determine any differences in length by checking for interference—if the waves arrive out of phase, they must have traveled different distances.

Astronomers employ interferometry to combine the output of several telescopes to achieve a higher resolution. In the process, the resulting image mimics that which would have been obtained from a much larger telescope. The Very Large Array (VLA), a set of 27 radio telescopes arranged in the shape of a *Y* near Socorro, New Mexico, achieves the resolution of an antenna that is 22 miles (36 km) across. Interfer-

Some of the radio telescopes of the Very Large Array *(Eastcott-Momatiuk/The Image Works)*

ometers that operate on much shorter wavelengths such as visible light instead of radio waves require a much greater precision in the positioning of the instruments.

Scientists and engineers at the Jet Propulsion Laboratory are working on a project to employ interferometry with small, space-based telescopes. This space interferometry mission, currently called SIM Lite, would improve stellar measurements by a factor of several hundred. Such precision would allow astronomers to image extrasolar Earths and even decide whether these planets are able to sustain terrestrial-like organisms. No launch date has been set for this project.

ESA has a similar mission on the drawing board. This project, called Darwin, would engage several space telescopes to search for and study extrasolar Earths. The telescopes would use interferometry to cancel the bright light of the star, thereby enhancing the capability of spotting any faintly illuminated orbiting planets.

The search for extrasolar planets of all varieties, especially those with Earth-like characteristics, will continue to be an exciting frontier of space and astronomy. This research can tell astronomers much about Earth and the solar system, as well as reveal the abundance—or not—of life in the galaxy. Some day soon astronomers may find life and possibly even other civilizations on faraway planets; or perhaps the search will come up empty, demonstrating the scarcity of life. Either way, it will be a profound and far-reaching discovery.

CHRONOLOGY

1543	The heliocentric theory, in which Polish scientist Nicolaus Copernicus (1473–1543) proposed that Earth and the other planets revolve around the Sun, is published.
1600	Church authorities burn Italian philosopher Giordano Bruno (1548–1600) at the stake for heresy. Among the beliefs that Bruno promoted was that the universe is infinite and populated with numerous worlds.
1610	Italian scientist Galileo Galilei (1564–1642) aims a telescope at Jupiter and finds some of its satellites.
1842	Austrian physicist and mathematician Christian Doppler (1803–1853) explains the frequency shift due to relative motion between the source of the wave and receiver.
1898	Early science fiction stories, such as H. G. Wells's *War of the Worlds,* sparks interest in extraterrestrial life.
1957	USSR launches *Sputnik,* the first artificial satellite.
1958	The United States government establishes the National Aeronautics and Space Administration (NASA).
1960s	Dutch-American astronomer Peter van de Kamp (1901–95) claims to have observe a wobble in the motion of Barnard's Star, which would suggest the

presence of planets orbiting this star, but this finding was not confirmed.

1963   Arecibo Observatory opens.

1980s   Canadian astronomers Bruce Campbell and Gordon Walker develop sensitive techniques to measure radial velocity.

1990   NASA's space shuttle *Discovery* lifts the *Hubble Space Telescope* into orbit.

1991   Pennsylvania State University astronomer Alex Wolszczan and his colleagues use the Arecibo Observatory to find planets orbiting a pulsar.

1993   Astronauts aboard the space shuttle *Endeavour* install a corrective optics package to fix a flaw in the optics of the *Hubble Space Telescope.*

1995   Swiss astronomers Michel Mayor and Didier Queloz at the Geneva Observatory use the radial velocity method to discover a planet orbiting a main-sequence star.

2006   European space scientists, led by the French space agency CNES (Centre National d'Études Spatiales—National Center for Space Research), launch COROT (Convection Rotation and Planetary Transits).

2009   NASA launches *Kepler* to search for Earth-like extrasolar planets.

FURTHER RESOURCES
Print and Internet

Bally, John, and Bo Reipurth. *The Birth of Stars and Planets.* Cambridge: Cambridge University Press, 2006. Beautiful photographs and illustrations reinforce the authors' explanation of current theories of star and planet formation.

European Space Agency. "COROT Discovers Smallest Exoplanet Yet, With a Surface To Walk On." News release, February 3, 2009. Available online. URL: http://www.esa.int/esaCP/SEM7G6XPXPF_Expanding_0.html. Accessed July 27, 2009. Astronomers announce finding an extrasolar planet about half the size of Earth with an orbital period of 20 hours.

Hubble Information Centre. "Hubble Finds First Organic Molecule on Extrasolar Planet." News release, March 19, 2008. Available online. URL: http://www.spacetelescope.org/news/html/heic0807.html. Accessed July 27, 2009. Researchers announce the discovery of methane in a Jupiter-sized extrasolar planet.

Miller, Ron. *Extrasolar Planets.* Minneapolis: Twenty-First Century Books, 2002. Aimed at students in grades 7–12, this book explains how astronomers discovered the planets in the solar system, and then extended their range beyond the local group.

National Aeronautics and Space Administration. "Hubble Finds Carbon Dioxide on an Extrasolar Planet." News release, December 9, 2008. Available online. URL: http://hubblesite.org/newscenter/archive/releases/2008/41/text/. Accessed July 27, 2009. Astronomers using the *Hubble Space Telescope* find evidence of carbon dioxide in the atmosphere of a hot Jupiter.

———. "Overview of the *Keppler Mission.*" Available online. URL: http://kepler.nasa.gov/about/. Accessed July 27, 2009. NASA discusses the importance of this mission, the scientific objectives, the methodology and design, and the expected results.

Space Telescope Science Institute. "Hubble Makes First Direct Measurements of Atmosphere on World around Another Star." News release, November 27, 2001. Available online. URL: http://hubblesite.org/newscenter/archive/releases/2001/38/text/. Accessed July 27, 2009. Astronomers use the *Hubble Space Telescope* to study light filtering through an extrasolar planet's atmosphere.

Thommes, Edward W., Soko Matsumura, and Frederic A. Rasio. "Gas Disks to Gas Giants: Simulating the Birth of Planetary Systems." *Science* 321 (August 8, 2008): 814–817. Researchers use computer simulations to study the formation of planets and find on many occasions that the results produce features similar to those that have been discovered in extrasolar planetary systems.

Webb, Stephen. *If the Universe Is Teeming with Aliens . . . Where Is Everybody? Fifty Solutions to Fermi's Paradox and the Problem of Extraterrestrial Life.* New York: Copernicus Books, 2002. The Italian-American physicist Enrico Fermi (1901–54) once posed the following question—if there are many civilizations in the galaxy, where are they? This book formulates possible answers to this fascinating question.

Web Sites

Extrasolar Planets Encyclopaedia. Available online. URL: http://exoplanet.eu/. Accessed July 27, 2009. Established in February 1995, this site compiles an interactive catalog of extrasolar planets that have been discovered so far.

NASA Exoplanet Science Institute. Available online. URL: http://nexsci.caltech.edu/. Accessed July 27, 2009. Formerly known as the Michelson Science Center, the NASA Exoplanet Science Institute is the operations and analysis center for NASA's extrasolar planet program. The Web site describes current missions and instruments, conferences, and data.

PlanetQuest: Exoplanet Exploration. Available online. URL: http://planet quest.jpl.nasa.gov/. Accessed July 27, 2009. NASA's Jet Propulsion Laboratory sponsors this Web site, which maintains the current planet count, and offers information on the science and technology of extrasolar planet searches.

COLONIZATION IN SPACE AND ON OTHER WORLDS

Earth provides an environment in which life can thrive, but it is not the only possible habitat. Astronauts have visited the Moon and spent weeks or months in orbiting spacecraft or space stations. The universe offers a vast amount of "elbow room." Earth is just one small body in a large solar system, which comprises only one star out of several hundred billion in the galaxy, and the galaxy is only one of billions of galaxies in the universe. Space is often referred to as the final frontier since it is the last to be explored, and much of it is still terra incognita—unknown territory.

People who argue for the establishment of a permanent presence beyond Earth can point to plenty of reasons. In addition to the adventure and freedom that frontier inhabitants enjoy, the expansion of living space would relieve the pressure of a burgeoning population on Earth. The increasing population is depleting Earth's resources as well as polluting the planet, perhaps irreversibly. Economic opportunities and an almost limitless supply of resources await beyond the confines of the smallish third planet from the Sun.

Population control and technological developments may be able to relieve the pressure on the planet's resources, but even so, the survival of human civilization is precarious if people do not venture beyond Earth. The fossil record shows that many species have come and gone, and episodes of mass extinction have periodically occurred, where many or even most species suddenly died out. Mass extinctions are probably due to cataclysmic events such as asteroid impacts or some other drastic and abrupt change. In an interview published in the *Washington Post* on September 25, 2005, Michael D. Griffin, NASA administrator for 2005–09, noted, "In the long run a single-planet species will not survive." As the old saying goes, people should be careful about putting all their eggs in one basket.

But space exploration and colonization poses serious risks. No human-friendly environments exist in the solar system except for Earth. Colonists can live in specially constructed habits drifting in space or orbiting a planet, or colonists can choose the firm ground of a satellite or planet, but whatever the choice, people who want to reside beyond Earth must take the essential aspects of their terrestrial environment with them.

A colony should be self-supporting and independent. This does not mean that colonists should have no contact with Earth, but they should be able to fend for themselves. If they cannot, the colony will drain the planet's resources and be just as subject to planet-wide catastrophes as an Earth dweller. This chapter discusses the science and technology of human life in space habitats or on moons or other planets in the solar system.

INTRODUCTION

The age of space exploration began on October 4, 1957, when the former Union of Soviet Socialist Republics (USSR) launched *Sputnik* (a Russian word for traveling companion), the first artificial satellite. On July 20, 1969, two United States astronauts, Neil Armstrong (1930–) and Edwin "Buzz" Aldrin (1930–), landed on the Moon. A total of 12 astronauts have walked on the Moon. The National Aeronautics and Space Administration (NASA) has also launched many probes to other planets, including all of the other rocky planets of the inner solar system—Mercury, Venus, and Mars. (See the sidebar on page 16 for more information on NASA.)

The first space station was USSR's *Salyut 1,* launched April 19, 1971. (*salyut* is Russian for "salute.") With a living space of about 3,500 feet^3 (100 m^3), it housed three crew members for about 24 days. Engineers did not design this station to be permanent, but rather as an experiment to begin testing the effects of long-term stays in space. Other space stations have also been built. The list includes additional stations in the Salyut program, the United States space station *Skylab,* USSR's *Mir* (a Russian word for peace or world), and the *International Space Station (ISS). ISS* is the only one that remains, and it will be described in the following section. The longest any person has stayed continuously in space is 438 days, a record that Russian cosmonaut Valeri Polyakov established when he came to *Mir* on January 8, 1994, and left on March 22, 1995. Polyakov flew around the world more than 7,000 times and traveled about 187,000,000 miles (300,000,000 km) in that time.

Another important spacecraft is NASA's space shuttle. After years of testing and development, the first space shuttle mission launched on April 12, 1981, and the shuttle continues to fly, although it is scheduled to be retired in 2010 (but may fly longer). Missions for the vehicle include the delivery of payloads such as satellites into space, a platform for scientific research in space, and as one of the primary means of ferrying astronauts and supplies to space stations. Unlike a one-shot rocket, the space shuttle is reusable, except for certain parts such as the external tank that are lost shortly after launch and must be replaced for the next mission. Three shuttles—*Discovery, Atlantis,* and *Endeavour*—compose the fleet as of 2009. Two vehicles have been lost in accidents: *Challenger* on January 28, 1986, and *Columbia* on February 1, 2003. Fourteen lives were lost in these tragedies.

An extremely important concept concerning satellites, whether natural ones such as the Moon or artificial ones such as space stations, is the notion of an orbit. The gravitational attraction between two masses is proportional to the product of their masses and inversely proportional to the square of the distance between them. Earth attracts any nearby object, and although the other object attracts Earth, if the planet has much more mass then it moves little and the other object moves a lot. Objects fall due to gravity, and this is also true of satellites. Although Earth's gravity decreases with the square of distance, it is still about 90 percent as strong at the altitude that the space shuttle typically orbits—about 200 miles (320 km)—than at the surface.

An orbiting object must travel at a certain, high rate of speed. What happens in an orbit is that the orbiting object falls, but it is going so fast that instead of hitting the ground, it sweeps around the planet. The figure illustrates the idea. No energy is required to keep an object in orbit, as long as it maintains its speed. But if an object loses speed, it will eventually fall; this means that orbits must be above the bulk of Earth's atmosphere, otherwise air resistance would cause the orbiting object to drop. Orbital speed depends on altitude. At the space shuttle's typical altitude, the craft must travel about 17,500 miles/hour (28,000 km/hr). Its orbital period is about 90 minutes—every hour and a half it completes a revolution around the globe.

Orbiting astronauts experience "weightlessness" or "zero gravity," so-called even though gravity is present (an orbit would be impossible without it, as explained above). The reason gravity seems to have disappeared is

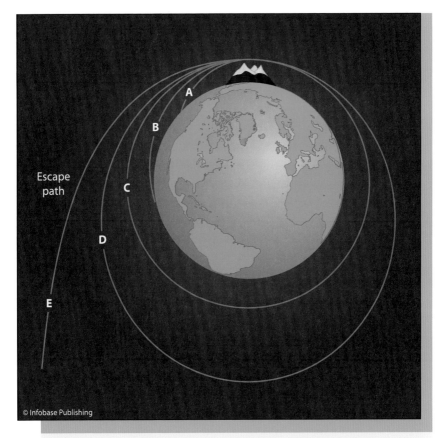

© Infobase Publishing

Suppose an object is thrown from a mountain. Gravity causes it to fall toward Earth, and if the velocity is too low, as in A and B, the object hits the planet. But greater velocities cause the object to "miss," continually falling as it moves around the planet, as in C and D. These objects are in orbit. At an even greater velocity, the object escapes Earth's gravity, as in E.

that the astronauts are falling along with the craft. The mass of the orbiting objects remains, but weight, which is the force of gravity acting on mass, is not noticeable since everything is falling together. The same effect would be observed if a falling person "drops" a key, which would seem to float in space because it is falling along with the person (neglecting air resistance). Since gravity is present, scientists often refer to "zero gravity" or "weightlessness" as *microgravity*.

The human body is adapted to life on Earth and its particular gravity. Earth's gravity at the surface is referred to as 1 G, where G stands for gravity. Astronauts who spend long periods of time in microgravity suffer a variety of physiological effects. A lack of normal gravity causes a shift in body fluids, impairs balance, and reduces blood cells. Although these problems quickly disappear when the astronaut returns to Earth, more serious situations may take much longer to correct. Muscle tissue diminishes—astronauts no longer have to resist the force of gravity—and people lose about 1 percent of their bone mass every month they live in microgravity. These losses cause dangerous weakness upon return to Earth until the body can recover.

Microgravity also requires a person to develop a new set of motor skills. Floating around a space station may seem like fun, but Newton's third law is at work—for every action there is an equal and opposite reaction. An astronaut may push off a wall to get to the other side, but if the push is too strong, a painful collision with the other wall results. Other tasks, such as drinking, eating, bathing, and using the bathroom also require special care.

Astronauts who visited the Moon, and those who will one day reach Mars or other planets, must deal with gravity at a different strength than that to which they are accustomed. The Moon has about a sixth of Earth's gravity, which meant that the equipment the lunar astronauts carried did not weigh so much, but Armstrong, Aldrin, and the others who walked on the Moon had to be careful they did not jump too high—and crash-land a great distance away! Mars has 0.38 G (38 percent of Earth's gravity).

Another danger of space exploration is the lack of air. Space is not a perfect vacuum—a few atoms and molecules are present—but it is close to one. In her book *Life at the Extremes,* Frances Ashcroft, a physiologist at Oxford University, described the unfortunate consequences of exposure to space without protective equipment: "Step into the vacuum of space and you would perish in a few brief, agonizing moments. The air would rush out of your lungs; the dissolved gases in your blood and body fluids would vaporize, forcing apart your cells and forming bubbles in your capillaries, so that no oxygen would reach your brain; air trapped in internal organs would expand, rupturing your gut and eardrums; and the intense cold would cause instant freezing."

Tragic accidents have occurred in space. USSR's *Soyuz 11* (*soyuz* is a Russian word for "union"), while returning from *Salyut 1* on June 30, 1971, developed a leak. Three cosmonauts on board died from depressurization.

The Moon also lacks an atmosphere, so astronauts and future colonists must observe the same precautions as in space. Mars has only an extremely tenuous atmosphere consisting mostly of carbon dioxide. The atmospheric pressure on Mars is less than 1 percent of Earth's.

Another serious worry is radiation. Earth's atmosphere absorbs or reflects most of the dangerous, high-frequency radiation and high-energy particles coming from space. Although some ultraviolet radiation from the Sun gets through—and is responsible for sunburns—much of this radiation, along with X-rays and cosmic rays, is blocked. Prolonged exposure to radiation generates serious health problems because the radiation damages vital biological molecules such as deoxyribonucleic acid (DNA) and kills cells, resulting in rapid death at high doses and potentially lethal diseases such as cancer at lower doses.

Spacecraft and spacesuits provide shielding to a limited extent, but astronauts carefully monitor radiation dosages, similar to workers at nuclear power plants. In addition, occasional intensification of radiation occurs during a solar flare—unusual activity in the Sun—which creates special hazards. Shielding that is thick enough to protect astronauts from these rare events is impractically heavy and cumbersome. Instead of increasing the shields, scientists monitor the Sun in order to forecast these events in advance, giving any astronaut time to seek shelter or return to Earth. But permanent colonies could gradually build sufficiently shielded emergency shelters.

INTERNATIONAL SPACE STATION (ISS)

The largest artificial satellite orbiting Earth is currently *ISS*. Construction began in 1998 and continues today, a little piece at a time, in orbit.

The *International Space Station,* photographed in 2005 from the space shuttle *Discovery,* with the Caspian Sea in the background *(STS-114 Crew/NASA)*

ISS orbits at an altitude of 240 miles (390 km). Visits from the space shuttle and other vehicles supply the crew and bring additional material and equipment for the next phase of construction. Enough of the craft was finished by November 2, 2000, to allow the initial crew—Expedition 1—to stay there for about four and a half months. Astronauts have been living at the *ISS* ever since, rotating every six months or so. Officials expect the station to be completed by 2011.

Because construction has taken place in stages, *ISS* is modular, composed of compartments or modules. Habitable space in the *ISS* has grown to the size of a typical three-bedroom house. Such an enormous volume would have been far too large for any existing launch system to lift all at once, which necessitated the piecemeal construction plan. The completed structure will exceed the volume of a five-bedroom house. *ISS* accommodates three crew members, but the final version will eventually house six. Because of its great size, people on the ground can sometimes see *ISS* with binoculars or even the unaided eye as it passes overhead.

ISS merits the term *international* in its name. Agencies involved in the construction and operation include NASA, European Space Agency (ESA), Russian space agency Roscosmos, Canadian Space Agency (CSA), Japan Aerospace Exploration Agency (JAXA), and others. Contributions have come from 16 nations: Belgium, Brazil, Canada, Denmark, France, Germany, Italy, Japan, the Netherlands, Norway, Russia, Spain, Sweden, Switzerland, United States, and the United Kingdom.

A source of energy is imperative to operate the life-support equipment that maintains and circulates the station's air supply, as well as to control devices and scientific instruments. Space stations could possibly rely on fuel brought up from Earth, but this would be expensive and risk disaster if the supply was delayed for any length of time. The safest solution and the best path for eventual success in space colonization is to be self-sufficient as much as possible.

ISS derives its energy from the Sun. Extending from the station like wings, solar panels absorb sunlight and convert the energy into electricity. The main arrays are rectangular, extending 238 feet (73 m) by 38 feet (12 m)—longer than the wingspan of a Boeing 777. When *ISS* is complete, the station will have about an acre of solar panels—27,000 feet^3 (2,500 m^3)—eight miles (12.8 km) of wire, and an electrical power capacity of 110 kilowatts.

This space station fulfills several purposes. As stated in NASA's fact sheet, "The *International Space Station* is vital to human exploration. It's where we're learning how to combat the physiological effects of being in space for long periods. It's our test bed for technologies and our decision-making processes when things go as planned and when they don't. It's important to learn and test these things 240 miles up rather than encountering them 240,000 miles away while on the way to Mars or beyond."

Science is also a critical part of the *ISS* mission. The orbiting station gives scientists an extended opportunity to study processes in microgravity. For example, crew members perform experiments in biology, crystallography, fluid physics, and much else. The United States designated the American segment of the *ISS* as a national laboratory in 2005.

FUTURE HABITATS

Lessons learned with *ISS* will be applied to future habitats beyond Earth, should they eventually be funded. Some important issues have already emerged.

Two potential sites for permanent colonies or habitats are orbiting stations, similar to *ISS,* and cities or shelters built on other worlds such as the Moon or Mars. The advantages and disadvantages of each of these options should be considered before a decision is made to pursue one or the other, or both.

People living in a space colony have the same basic needs as people living on Earth—food, air, some degree of privacy, energy, transportation, communication, materials for construction, and protection from solar and cosmic sources of radiation. The advantage of living in a habitat orbiting Earth is that it is close to home. Proximal colonies are easier to get started and less costly to keep supplied until the colonists can begin to rely on themselves. Complete self-sufficiency is highly unlikely for a considerable period of time, so a short supply route reduces start-up expenses. And in an emergency, Earth is available for a quick escape.

Orbital colonies also have enormous potential for growth. In space, colonists need not seek and defend a patch of ground upon which to build their home because they can construct one virtually anywhere. If Earth's orbital area begins to get crowded, colonists can build habitats orbiting the Sun or other worlds. Although the increased distance from the mother planet would obviate some of the advantages of orbital colonies, by the time space around Earth gets crowded, people will have probably developed a potent space-based economy and transportation system. Orbital colonies enjoy virtually unlimited potential.

One of the main disadvantages is the absence of any handy source of material. People living "dirt side"—on a world—can go outside and scrounge up what they need. For people living in an orbital colony, the outside is a vacuum. Perhaps if space becomes densely populated, materials and supplies will be readily available from orbiting depots or supply ships. But in the beginning, orbital colonies must engage in the expensive operation of transporting materials from Earth.

The problem is the strength of Earth's gravity. In the simplest formulation of Newton's second law—a law of motion discovered by the British physicist Sir Isaac Newton (1642–1727)—acceleration of an object equals the force applied to it divided by its mass. To counter Earth's

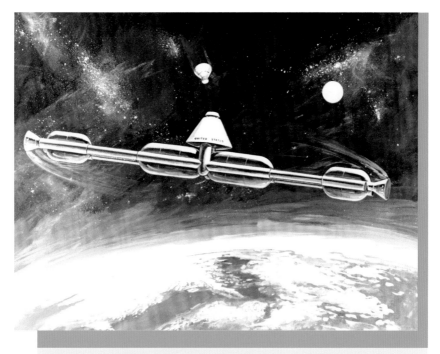

In this early 1960s design for a space station, rotation along the central axis simulated gravity. *(NASA)*

gravitational force, which exerts a strong pull, a climbing rocket needs a lot of power. From a starting point on the surface of the planet, escaping Earth's gravity and venturing out into space requires a speed of about seven miles/second (11.2 km/s)—about 25,000 miles/hour (40,000 km/hr)! This velocity is called the *escape velocity*. Even attaining an orbit at an altitude typical of the space shuttle or *ISS* requires a speed of about 17,500 miles/hour (28,000 km/hr). Such tremendous speeds require special engines and sturdy vehicles, as discussed in chapter 3. This is the reason why the space age did not begin until 1957.

Typical space launches cost millions of dollars. The price per pound is generally around $5,000–$10,000 ($11,000–$22,000/kg), depending on the launch vehicle. As Frances Ashcroft noted in her book, "The Apollo missions to the moon cost the United States a massive 4.5 per cent of its annual budget." Billions have been spent on *ISS,* a lot of which has been spent boosting personnel and material into space. These costs hold for space launches to other worlds as well as orbital space stations, but orbital colonies might need continual service to supply raw materials.

Artificial Gravity

Space travelers on science fiction television programs often walk around on their ship, clearly in the presence of gravity. Some of these programs make vague references to "gravity machines" that are capable of generating this force. Perhaps such technology may be developed in the future, but unfortunately for NASA and space programs all over the world, it does not exist today.

Gravity can be simulated with acceleration, either linear or angular. (It was this observation, among others, that led the German-American physicist Albert Einstein [1879–1955] to the general theory of relativity, the theory that physicists use today to understand gravity and many aspects of cosmology.) Linear acceleration—changing forward speed—results in the familiar push in the back that car and airplane passengers feel as the vehicle picks up speed. This is not a practical way of simulating gravity in space because it requires energy to continue accelerating. An object on a rotating cylinder, however, is continually pulled outward, an effect known as centrifugal force. This force is due to the object's inertia—objects tend to travel in a straight line. The rotation of the craft keeps the object spinning, as shown in the figure. As a result, objects within the hull of a spinning ship experience "gravity." A person standing on the inside of the hull feels weight, for example, and a dropped object "falls" to the floor.

Rotation is a practical means of achieving artificial gravity because a space station will keep spinning without an additional energy input once it has been set into motion, due to a lack of friction or air resistance in the vacuum of space. The rotation rate depends on the size of the habitat and the desired "gravity." If the habitat is too small or spinning too rapidly, residents would become nauseated. To achieve 1 G

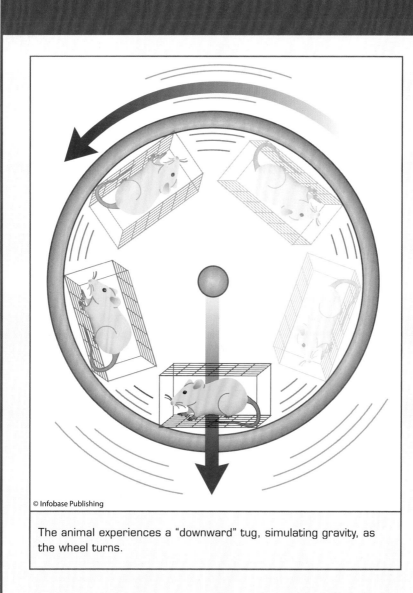

© Infobase Publishing

The animal experiences a "downward" tug, simulating gravity, as the wheel turns.

at one revolution per minute, the craft would need to have a diameter of 5,900 feet (1,800 m)—quite large. If the inhabitants could tolerate less "gravity" or a higher rotation rate, the habitat could be much smaller.

Another disadvantage of an orbital colony is microgravity. Astronauts are well trained and physically fit, but they still encounter problems orienting to microgravity. Space sickness is common. This malaise appears to be similar to sea sickness and car sickness, resulting from confusing or mismatched sensations; certainly the sight of a person floating around the cabin of a space ship, and even "standing" upside down, would be disconcerting.

Space sickness usually disappears shortly, but the bone and muscle tissue losses are more serious. Perhaps these losses would not disturb orbital colonists who have no plans to return to Earth's strong gravity. If colonists do need strong bones and muscles to withstand gravity, perhaps they can make do with less tissue. But the physiological consequences of long-term microgravity on the human body are not understood. This gap in knowledge is unlikely to change until space pioneers begin living on the first permanent orbital stations.

An alternative is for orbital colonists to bring gravity with them. Gravity can be mimicked or simulated by rotating the station. This force is what keeps riders doing the loop-the-loop—a vertical loop—in their seats during the upside down portion of the path. Simulated gravity is often referred to as *artificial gravity*. The sidebar on page 40 explores this topic in more detail.

LIFE ON THE FRONTIER

Life on an orbiting space station or on a small world such as the Moon or Mars will require substantial adjustments. Gravity is likely to be only a fraction of the customary strength. Colonists must learn to adapt.

New environments bring challenges and obstacles, as well as potential benefits. Microgravity might be preferable over high-gravity environments because it could be easier on the heart and bones. Heart disease is the leading cause of death in the United States, and people with weak hearts may go to an orbital colony so that the cardiac muscle does not have to work against gravity.

One way that astronauts have been trying to maintain their health in space is by exercising. NASA astronaut Edward Lu lived at *ISS* from April 25, 2003, to October 27, 2003, as a science officer. Lu wrote about his experience in a column, "Greetings, Earthlings," at NASA's Web site. "We rotate our workouts through four main pieces of exercise equipment on *ISS:* a treadmill, 2 stationary bikes, and an apparatus that allows you to 'lift weights.'" Exercise in microgravity might seem odd. Lu explained that the treadmill was a "monster." "The treadmill is so big because it isn't bolted to the floor, but rather is loosely suspended inside a pit in the floor and has a big gyroscope inside that stabilizes it while you run. This is to isolate the vibrations from your footsteps so they don't shake the Station around. In effect, you are running on a floating treadmill." In order to run in microgravity, "we wear a harness (that fits like a backpack harness) that is connected to the treadmill with bungee straps."

Lifting weights in a "weightless" environment also requires some adaptation. Lu wrote, "The piece of equipment that I use the most goes by the acronym RED (resistive exercise device), which is our version of lifting weights. It is two canisters, each about the size of a watermelon, that are bolted to an aluminum plate. Inside each canister is a cord that wraps around a spiral pulley that unwinds as you pull the cord out. The resistance is provided by a stacked series of disks with rubber spokes inside the canister connected to the spiral pulley. The whole assembly inside the canister winds up like a rubber band-powered toy as you pull the cord. You choose the resistance level by turning a crank that winds the disks in the opposite direction."

Exercise helps minimize the muscle losses experienced in microgravity, at least over the durations of most *ISS* tours of duty, but the loss of bone tissue continues to be a problem. No one is certain if durations of indefinite length in microgravity are possible.

Space colonists must also endure isolation. People often spend vacations in isolated spots so that they can "get away from it all," but many colonists will be making this choice permanently—going back is an expensive option, and one that may have limited availability. Such a decision requires a certain pioneering spirit. But getting along with other spirited pioneers is also imperative; enclosed in a risky and delicate habitat with nowhere else to go, colonists must learn to solve conflicts with tact and compromise rather letting arguments fester. Isolation similar to that encountered in space occurs on Earth at polar research stations and nuclear submarines, where personnel must display their interpersonal skills. But these situations do not quite approach the magnitude of isolation in space. Colonists must have an understanding of psychology as well as physiology in order to adapt to life in space.

How the colonists organize and govern themselves will depend a lot on how the first colonies are funded. Governments have funded

An early use of Biosphere 2, a large enclosure near Tucson, Arizona, was to simulate an isolated space colony. Researchers now employ the structure in studies of ecosystems. *(Martin Bond/Photo Researchers, Inc.)*

most space exploration to date since large sums of money have been required. Should this continue, terrestrial government involvement in the colonies might be expected—which could have its drawbacks, considering the distance separating a government on Earth and a space colony. Some organizations such as the Space Frontier Foundation encourage private investment in space activities. The following sidebar provides more information on this space advocacy group.

The private sector is beginning to get involved. On October 4, 2004, SpaceShipOne, built by the company Scaled Composites, became the first manned nongovernment spacecraft in space. On this day, the vehicle achieved an altitude of 328,000 feet (100,000 m) for the second time in a two-week period.

Space Frontier Foundation

Activists such as Rick Tumlinson and others founded the Space Frontier Foundation in 1988. The power of activism comes in getting the message out and spreading a particular point of view. In the case of the Space Frontier Foundation, the message is that space is a frontier open to all. One of the first acts of the foundation was to market physicist Gerard K. O'Neill's book *The High Frontier,* which describes the colonization of space and the development of space industrialization—the opportunities for industry in space.

One of the foundation's current projects is to foster space-based solar power. People and businesses sometimes use solar energy to heat buildings and generate electricity, which produces much less pollution than burning oil or coal, though solar energy tends to be more expensive. *ISS* and many other spacecraft use solar energy, and there is plenty of sunlight in space—never a cloudy day. The Space Frontier Foundation is encouraging businesses to make investments and develop technology to create a power utility that would capture solar energy in space and beam it to customers on the ground. This utility would be a wise use of resources as well as promoting the industrial development of space.

Another project, jointly operated with the United States Rocket Academy, is Teachers in Space. The goal is to train teachers for brief journeys into space. These teachers will then be able to share what they have learned with their students. In a news release issued by the Space Frontier Foundation on February 7, 2009, project manager Edward Wright said, "Astronaut teachers will fly during the summer and return to the classroom in the fall with a priceless payload of knowledge and firsthand experience. Every astronaut teacher will reach and inspire hundreds of students every year. For the first time, space can have a real effect on American education."

Artist's rendition of a lunar colony *(Victor Habbick Visions/"Photo Researchers, Inc.)*

LUNAR COLONY

An alternative to orbital habitats is to colonize other worlds. The closest one is the Moon, which is about 240,000 miles (384,000 km) distant from Earth. Apollo astronauts who journeyed to the Moon took about three days to reach their destination (but they landed a day or so later).

The radius of the Moon is about a quarter of Earth's radius, and it has only 1/80th the mass. Although it may have had an atmosphere at one time, almost all of the gases have since escaped, so the Moon is virtually an airless world. The lack of atmosphere results in uneven temperatures since there is no air to distribute heat. On the sunny side, temperatures can hit 212°F (100°C), while the dark side is a frigid -283°F (-175°C). The surface is heavily cratered since falling rocks do not burn up before striking the Moon—no atmosphere to generate friction—and there is nothing to weather or erode the craters once formed. Lunar soil is fine-grained, consisting of many pieces of round, glassy particles. Walking on the Moon is a slippery and dusty adventure.

Ownership of orbiting space stations is obvious—whoever built and paid for the station is the owner—but the concept of property on other worlds invites tricky questions. There is an American flag on the Moon, planted by American astronauts, but the issue of ownership would entail fewer disputes if all interested parties signed a treaty to establish procedures for filing claims. One treaty, the 1979 Moon Agreement, proposes that all people must share the Moon's resources. But only about a dozen countries have ratified this treaty, and the United States, as well as most other space-faring nations, is not among them. This issue should ideally be solved prior to any serious attempt at colonizing the Moon.

On January 14, 2004, then president George W. Bush issued his vision for future space exploration. As discussed in the 2004 NASA document *The Vision for Space Exploration,* part of President Bush's plan was to return to the Moon with manned flights. One of the initiatives was as follows: "Conduct the first extended human expedition to the lunar surface as early as 2015, but no later than the year 2020." But subsequent administrations may not be as enthusiastic about space exploration, so these target dates are extremely uncertain.

The advantages of establishing a colony on the Moon include the presence of terra firma—solid land—and one-sixth of Earth's gravity is better than none. Lunar escape velocity is about 5,400 miles/hour (8,640 km/hr), much less than Earth's escape velocity, which facilitates transportation. Materials are available to build shelters.

But scientists are not sure that Moon's gravity is sufficient to prevent the serious bone losses suffered in microgravity. The perils of radiation and a vacuum would be present, and although the Moon is the closest major body, it is much farther than a habitat orbiting Earth. Another disadvantage is that while colonists could use lunar rocks and soil to build habitats, the Moon lacks elements such as carbon and nitrogen that are needed to make food, fuel, and other necessities. Oxygen abounds in the lunar crust, but it is locked up in minerals and may be hard to release. Food and water would have to be manufactured, as they would in an orbital colony.

There is also the problem of energy. Solar power can be used, but the lunar "night" is long—equivalent to about 14–15 days on Earth. This long dark period is because one side of the Moon always faces Earth; the Moon revolves about the planet once a month, and a point on the lunar surface faces away from the Sun during about half of this journey. Such a long time between sunlight taxes the ability of solar power systems to store energy.

But NASA is working on solving this problem. In a news release issued on September 10, 2008, NASA scientists announced plans to develop a system based on nuclear fission that can generate enough power for a small colony. According to Lee Mason, a scientist at NASA's Glenn Center in Cleveland and the lead researcher on the project, "Our long-term goal is to demonstrate technical readiness early in the next decade, when NASA is expected to decide on the type of power system to be used on the lunar surface." These and other problems must be solved before colonists venture to the Moon.

MARS COLONY

The next best bet for colonization of another world is Mars. Mars, the fourth planet from the Sun, orbits the Sun at an average distance of about 141,000,000 miles (228,000,000 km), or about 1.5 astronomical units. The radius of Mars is about half that of Earth, and the "Red Planet"—so called because it has a reddish surface due to iron oxide—has a mass of about a tenth that of Earth. At 0.38 G, Mars's gravity is slightly greater than a third of Earth's. In its closest approach, Mars gets within 35,000,000 miles (56,000,000 km) of Earth.

Despite its greater distance, Mars has some advantages over the Moon for prospective colonists. Gravity is stronger, which reduces the need for physiological adaptation. Unlike the Moon, Mars has a measurable atmosphere, which consists mostly of carbon dioxide with a little nitrogen and argon, and traces of other elements, including oxygen. This atmosphere offers some degree of protection from radiation, the dangers of which are further weakened by the planet's greater distance from the Sun.

Mars and Earth have a few other properties in common as well. Because Mars revolves at a similar rate to Earth, the Martian day is only slightly different, equal to about 24 hours and 40 minutes. The planet's tilt is also remarkably similar to Earth, which results in seasons such as summer and winter, but these seasons last a lot longer because the Martian year is almost twice as long.

Perhaps the most significant advantage that Mars offers is the presence of water in the form of ice. Surface features such as channels, mapped by orbiting satellites, indicate that water once flowed on Mars, a finding confirmed by space probes that have landed on and roamed about the planet. Although these channels are now dry, Mars possesses polar ice caps that contain some frozen

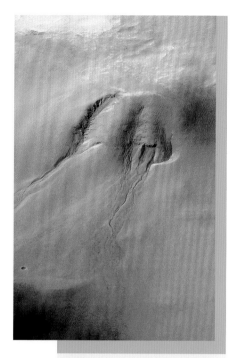

Taken from the *Mars Global Surveyor,* this photograph of the surface of Mars shows channels that may have carried water during warmer periods in the planet's history. *(NASA/Jet Propulsion Laboratory)*

water, along with frozen carbon dioxide (dry ice). NASA scientists estimate that the ice caps contain enough water to cover the planet to a depth of about 33 feet (10 m). There is also probably ice underneath the surface, and perhaps even flowing water.

The presence of water and other substances on Mars at least partially offsets the disadvantage of its greater distance from Earth. In its beginning phases, the first colony on a world will probably need a steady train of supplies from Earth, which will be more time consuming and expensive to deliver to Mars than to the Moon. But a Martian colony is likely to be up and going more quickly because so many of the essential raw materials are at hand.

Despite the advantages, Mars is not exactly hospitable. The average temperature is about ⁻76°F (⁻60°C)—colder than Antarctica. Although Mars has an atmosphere, the pressure is only about 1 per cent that of Earth. This pressure is not enough sustain life, so humans would need to wear pressurized suits outside of their habitat. Mars also does not have an appreciable magnetic field to deflect high-speed *ions* coming from space.

The long travel time to the planet will also be a problem. Engines that power satellite-launching rockets as well as the space shuttle rely on large amounts of fuel, which greatly increase liftoff weight. To cover long distances, space probes conserve fuel by entering orbits around the Sun, and "falling" toward the destination. These trajectories extend the journey considerably, but a direct route would require an unpractical amount of fuel. Colonists traveling to Mars can expect to spend about nine months in space if they use the standard rockets of today. Unless the ship generates artificial gravity—and it would likely be too small to do so safely—the colonists could arrive at their destination in an unhealthy condition, with bones too brittle to get much work done, even though Mars has a gentler gravity. And the decreased gravity may limit recovery. The only other option would be to develop a faster means of transportation, such as a nuclear-powered spaceship. Chapter 3 discusses advances in propulsion science.

One of the biggest advantages of Mars might also be a drawback. Mars resembles Earth in many ways, causing some people to wonder if it developed life at some point in its past. Perhaps microscopic life forms persist, buried under the surface, although there is currently no evidence for this. If an unmanned probe finds life on Mars before any colonists arrive, the discovery would raise questions about possible contamination and diseases, as well as ethical concerns about "invading" another life-form's domain.

TERRAFORMING—PLANETARY MAKEOVER

One way to increase the odds of successful settlement on Mars is to make the planet more hospitable. This concept goes by the name of *terraforming*—the shaping or formation of Earth-like environments.

Mars would be an excellent terraforming candidate, providing that it harbors no life. Since terraforming could wipe out native Martian organisms, the presence of life on Mars, even single-celled microbes, would probably raise too many ethical and scientific issues for terraforming to meet with general approval.

Assuming Mars is currently sterile, how could people engineer the planet's environment? To make a planet more livable, the environment must provide warmth, water, air, food, and fuel. Mars already has a supply of water, though it is frozen; if engineers could heat the planet a little, the ice would melt. As for fuel, most of the energy on Earth derives from the Sun, and although fewer of the Sun's rays strike the more distant Mars, plenty of solar energy is available. Food sources such as plants can utilize this source if the atmosphere becomes thicker and temperatures rise.

Much would be accomplished if terraformers could add a little heat and air. A neat trick would be to do both at the same time. This trick might be feasible if the air contained a significant quantity of *greenhouse gases.*

Greenhouse gases in the atmosphere trap heat because they tend to absorb electromagnetic radiation in the infrared portion of the spectrum. Radiation is one of the three main ways that heat moves around, the other two are conduction and convection, which are less important in space because they need a solid or fluid medium in order to flow. Warm objects emit a lot of infrared radiation, which has a lower frequency (and therefore lower energy) than visible light. Greenhouse gases do not block visible light and the more energetic ultraviolet radiation from the Sun, and this radiation warms the planet and objects on its surface. But when the warmed ground cools off by emitting infrared radiation, greenhouse gases block the radiation from escaping into space. The extra heat warms the planet. Some of the gases that excel in this greenhouse effect are carbon dioxide, methane, and water.

Suppose engineers released a large amount of carbon dioxide on the surface of Mars, perhaps by building chemical factories. As the atmosphere became thicker, the trapped heat would warm the planet. Increased temperature and pressure would melt the ice, allowing water to flow on the surface. The introduction of certain species of hardy photosynthetic organisms such as algae would generate oxygen and further transform the atmosphere. These organisms could also become the basis of a food chain.

Scientists are not certain if this kind of terraforming process would work. In theory it might, but global climates are complicated, and any number of factors could intervene to cause the greenhouse gases to fly away into space, leaving frustrated planetary engineers with the same thin atmosphere as before. Even if successful, many years might have to pass before Mars became anywhere close to being habitable without special equipment and habitats.

CONCLUSION

In the relatively short period of its existence, the space-faring age has seen several dramatic achievements. Astronauts have journeyed to the Moon, walked around, and driven a lunar rover over its surface. Scientists and astronauts have spent considerable time in orbit, and the *International Space Station* continues to provide an orbiting environment where researchers study the effects of microgravity.

But so much more is possible. The solar system is rich in room and resources. Yet progress seems slow, at least to those who are impatient to get into space. Money is one of the primary issues—space launches are extremely expensive due to the many requirements, including fuel, specialized equipment, and the expertise of a highly trained staff.

Some of the wealthier space enthusiasts have already experienced a journey. In 2001, for example, the California businessman Dennis Tito paid the Russians to take him to *ISS* in a *Soyuz* spacecraft. (The amount of money Tito paid has not been disclosed, but was probably millions of dollars and might have been as much as $20 million.) Following the launch on April 28, 2001, the 60-year-old businessman joined astronauts at *ISS* for more than seven days. He returned home safely on May 6, 2001. Since then, several other space tourists have enjoyed trips to *ISS*.

Most people cannot afford to be a space tourist. Space enthusiasts without adequate financial resources must wait until the frontier more fully develops, so hopeful pioneers closely monitor the research and development projects described in this chapter.

Advances in space exploration will come, albeit slowly, if governments continue to invest some amount of money in agencies such as NASA, ESA, and others. The process could be accelerated with the inclusion of commercial enterprises, as promoted by the Space Frontier Foundation and similar advocacy groups. Stepping-stone projects include a larger orbital space station, a renewal of manned exploration of the Moon, and perhaps the establishment of a small lunar base. These projects, though only first steps, could lead to the development of space colonization.

But there is a possibility that humans may not be able to wait. Environmental concerns, global climate change, disastrous scenarios such as the approach of an asteroid on a collision course with Earth, and other threatening developments may make a sudden move imperative. If no hospitable worlds are available, many humans may need to venture into space, perhaps to orbital stations that would quickly become overcrowded.

Another possible option is an interstellar journey. As chapter 1 discussed, the solar system is not the only planetary system in the galaxy. Other worlds exist, although the distances are staggering. The nearest star system, Alpha Centauri, is 4.35 light-years away—25.6 trillion miles (41.2 trillion km). Such distances are far too great for humans to travel—the trip would take much longer than a human lifetime—unless certain advances in technology occur. Advances in spaceship and propulsion technology may shorten the journey, and if spaceships can attain speeds close to the speed of light in a vacuum—180,000 miles/second (300,000 km/s)—then biological clocks will slow down, as described in Einstein's relativity theory. Scientists may also learn how to extend the human life span by suspended animation.

But even in the absence of these technological breakthroughs, nothing prevents humans from setting out on a long journey providing they are willing to let their descendants finish it. The concept is called a *generation ship* or interstellar ark. Often conceived of as a huge version of a space station, the vehicle would not be orbiting a star or planet but would instead travel interstellar distances. Rotation would provide artificial gravity. The occupants must be totally self-sufficient, growing their own food and producing their own energy, perhaps by nuclear reactions. Parents would spend their lives aboard the habitat, raising children who would in turn produce the next generation, and so on, until the ship finally reaches its destination. The descendants of the original colonies might arrive at another world, perhaps to encounter other civilizations.

Assuming that human civilization endures, people will almost assuredly colonize space in one way or another. Perhaps the pioneers will succeed with one of the ideas described in this chapter, or perhaps future scientists or entrepreneurs will conceive of something better. In either case, people will one day reach far into this frontier as human society continues to expand and consume more resources than one planet is capable of giving.

CHRONOLOGY

1957	The former USSR launches *Sputnik,* the world's first artificial satellite.
1958	*Explorer 1,* the first successful American satellite, launches.
	The United States establishes NASA.
1959	USSR launches *Luna 2,* the first artificial object to reach the Moon's surface.
1961	USSR cosmonaut Yuri A. Gargarin (1934–68) becomes the first person in space, orbiting Earth in *Vostok 1.*
1965	USSR cosmonaut Alexei A. Leonov (1934–) performs the first spacewalk (venturing outside of the spacecraft).
1967	A fire in the command module of a spacecraft sitting at Cape Kennedy (later to be named Cape Canaveral) in Florida kills three American astronauts.
	USSR *Soyuz 1* crashes, killing cosmonaut Vladimir M. Komarov (1927–67), the first spaceflight casualty.
1969	American astronauts Neil Armstrong (1930–) and Edwin "Buzz" Aldrin (1930–) walk on the Moon.
1971	USSR launches *Salyut 1,* the first space station. It stays in orbit until 1973.

1972	American astronauts Harrison H. Schmitt (1935–) and Eugene A. Cernan (1934–) walk on the Moon, the last astronauts to have done so.
1973	Experimental space station *Skylab* launches.
1981	The initial manned flight of a space shuttle, *Columbia,* occurs.
1984	American astronaut Bruce McCandless (1937–) performs the first untethered spacewalk.
1986	The space shuttle *Challenger* explodes soon after lift-off, killing all seven crew members.
	USSR launches the space station *Mir.*
1998	Space shuttle *Discovery* lifts off with astronaut John Glenn (1921–). The 77-year-old Glenn becomes the oldest person in space.
	In-space assembly of *ISS* begins.
2000	The first crew settles into *ISS.*
2001	California businessman Dennis Tito becomes the first space tourist, spending a week aboard *ISS.*
2004	President George W. Bush announces a space initiative calling for a return of manned exploration of the Moon and then on to Mars.
2009	Additional solar panels are installed at *ISS.*

FURTHER RESOURCES

Print and Internet

Ashcroft, Frances. *Life at the Extremes.* Berkeley: University of California Press, 2000. Ashcroft discusses the biology of living in extreme environments and situations. Topics include mountaintops and other low-pressure locations, the deep sea, hot and cold spots, strenuous exercise, and life in space.

Hall, Theodore W. "Artificial Gravity and the Architecture of Orbital Habitats." Available online. URL: http://www.spacefuture.com/archive/artificial_gravity_and_the_architecture_of_orbital_habitats.shtml. Accessed July 27, 2009. This article reviews artificial gravity and how it may be incorporated into the designs of space stations.

Lee, Wayne. *To Rise from Earth,* 2nd ed. New York: Facts On File, 1999. This book offers an excellent discussion of space exploration.

Lu, Ed. "Greetings, Earthlings: Ed's Musings from Space. Working Out." Available online. URL: http://spaceflight.nasa.gov/station/crew/exp7/luletters/lu_letter7.html. Accessed July 27, 2009. *ISS* astronaut Ed Lu describes his workout regimen.

National Space and Aeronautics Administration. *Fact Sheet: International Space Station.* Available online. URL: http://spaceflight.nasa.gov/spacenews/factsheets/pdfs/iss_fact_sheet.pdf. Accessed July 27, 2009. This two-page file provides a concise description of the *ISS.*

———. "NASA Developing Fission Surface Power Technology." News release, September 10, 2008. Available online. URL: http://www.nasa.gov/home/hqnews/2008/sep/HQ_08-227_Moon_Power.html. Accessed July 27, 2009. NASA researchers announce plans to design a nuclear reactor to help supply power for lunar colonies.

———. *The Vision for Space Exploration.* Available online. URL: http://www.nasa.gov/pdf/55583main_vision_space_exploration2.pdf. Accessed July 27, 2009. NASA discusses the future of space exploration, including President Bush's 2004 initiatives.

O'Neill, Gerard K. *The High Frontier.* New York: Morrow, 1977. In this classic book, O'Neill, a physicist at Princeton University, discusses space exploration issues at the time when the Apollo program had just come to an end.

Reynolds, David West. *Apollo: The Epic Journey to the Moon.* New York: Harcourt, 2002. This beautifully illustrated book describes the Apollo voyages to the Moon.

Schmidt, Stanley, and Robert Zubrin, eds. *Islands in the Sky: Bold New Ideas for Colonizing Space.* New York: Wiley, 1996. The articles in this book flesh out many of the ideas at the cutting edge of space science, including terraforming, rocket propulsion, and interstellar commerce.

Schmitt, Harrison H. *Return to the Moon: Exploration, Enterprise, and Energy in the Human Settlement of Space.* New York: Copernicus Books, 2006. Schmitt is an astronaut who has walked on the Moon. In this book, he outlines the future of space exploration.

Space Frontier Foundation. "Teachers Help Design New Astronaut Curriculum." News release, February 7, 2009. Teachers met at the Johnson Space Center in Houston, Texas, in February to discuss the training of astronaut teachers.

Washington Post.com. "NASA's Griffin: 'Humans Will Colonize the Solar System.'" September 25, 2005. Available online. URL: http://www.washingtonpost.com/wp-dyn/content/article/2005/09/23/AR2005092301691.html. Accessed July 27, 2009. This article consists of an interview with Michael Griffin, NASA's administrator from 2005 to 2009, who discusses why manned space flights are essential.

Web Sites

National Space Society home page. Available online. URL: http://www.nss.org/. Accessed July 27, 2009. Dedicated to the vision of human colonization in space, the National Space Society's Web site describes the organization's activities and the future of space exploration.

Planetary Society home page. Available online. URL: http://www.planetary.org/. Accessed July 27, 2009. The Planetary Society is a large organization devoted to promoting space exploration and the search for extraterrestrial life. Their Web site describes the organization's functions and provides news of the latest space developments.

Space Frontier Foundation home page. Available online. URL: http://spacefrontier.org/. Accessed July 27, 2009. The Web site of the Space Frontier Foundation presents the organization's mission to foster the commercial development of space and details their history and projects.

3

TRAVELING AMONG THE STARS

Ships capable of zooming through the vast distances between the stars have long been a part of fascinating adventures in science fiction stories. On the television series *Star Trek,* Captain James T. Kirk and his crew rapidly covered interstellar distances on the *Enterprise,* powered by the fictional "warp drive."

As chapter 1 described, scientists have begun to discover extrasolar planetary systems, which provide incentive and destinations for interstellar travel, but the distances from Earth are immense. The closest star system, Alpha Centauri, is 4.35 light-years away—25.6 trillion miles (41.2 trillion km). A ship traveling at Earth's escape velocity of 25,000 miles/hour (40,000 km/hr), which rockets of NASA and other space exploration agencies have achieved, would take about 115,000 years to reach this star system. Manned exploration of the galaxy, which consists of billions of stars within a disk about 100,000 light-years in diameter, is out of the question with such spacecraft.

Yet researchers have not abandoned notions of interstellar travel. Although starship propulsion is beyond the present state of technology—except for large generation ships, in which many generations of the crew live and die before reaching the destination—advances in propulsion methods could result in practical interstellar travel, at least to the nearest stars. The rewards of reaching the stars would be many. In addition to the rich potential of colonization (see chapter 2), people would be able to gauge the plurality of worlds and discover how many of them harbor life. Humans

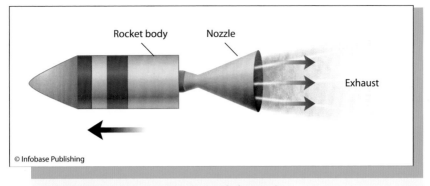

Rocket body   Nozzle

Exhaust

© Infobase Publishing

High-speed exhaust propels the craft forward.

would finally determine whether civilization and technological societies are common or rare.

This chapter discusses the rockets that have been built and launched, how they work, and how researchers are trying to improve them. Since manned interstellar travel involves the biological endurance and needs of humans, physiology is as important as physics. The chapter also covers advanced starship drives, and what is possible under the laws of physics as they are currently understood.

INTRODUCTION

The rockets that lift missiles, satellites, and the space shuttle use the principle of jet propulsion. Jet engines create thrust by emitting a stream of gas, generated by combustion or other chemical reactions. The backward stream of exhaust propels the vehicle forward due to *Newton's third law,* discovered by the British physicist Sir Isaac Newton (1642–1727), which says that every action has an equal and opposite reaction. Sending a jet of particles out of the back of the engine is the action, and the reaction is the forward motion of the craft, as illustrated in the figure.

Speed and mass of the emitted projectiles are critical to the engine's thrust—the force that propels it. The *momentum* of an object equals the product of its mass and velocity. In the absence of external forces, momentum is conserved. For a rocket, this means that the momentum of the expelled gas must be equal in magnitude to the momentum of the rocket but opposite in sign, so that the sum is zero. (This means there is no change in the momentum of the system—the engine and its exhaust. In other words, momentum is conserved.) This is the "equal and opposite" aspect of Newton's third law—the greater the momentum of the exhaust, the greater the momentum of the engine. Rockets can achieve substantial thrust by exhausting particles of small mass at extremely high speed, or particles of larger mass at lower speeds. In either case, the rocket shoots forward, picking up speed as its momentum increases to offset the momentum of the exhaust.

Rockets have long been made with a propellant of gunpowder, which made impressive displays or weapons. The modern development of rocket technology and associated propulsion techniques hinged upon improvements in design and fuel. In 1903, Konstantin Tsiolkovsky (1857–1935), a Russian mathematics teacher, published the basic formula of rocket flight and discussed the nature of propellants. The American physicist and inventor Robert H. Goddard (1882–1945) launched the first liquid-fueled rocket on March 16, 1926. A supply of liquid oxygen and gasoline powered the slender rocket to an altitude of only about 41 feet (12.5 m) in a flight that lasted 2.5 seconds.

Despite the inauspicious start, Goddard was on the right track. What rocket scientists needed was a fuel that would produce a large volume of hot, expanding gas. Most automobile engines also exert force by the expansion of hot gases, which act on pistons. In automobiles, combustion of gasoline, an energy-rich hydrocarbon fuel, creates the expanding gas. Combustion generally requires oxygen—the burning of a substance involves chemical reactions with oxygen, along with heat or a spark to get the process started. Automobiles can run on gasoline because they operate in Earth's atmosphere, which is about 21 percent oxygen. All that is needed is to draw a little air into the combustion chamber. Without air intake, an automobile engine fails; at high altitudes, where the atmospheric pressure is much less than sea level, automobile engines run poorly. But if rockets are to function in space, which contains an extremely low density of molecules, they must do without oxygen or carry it along with them.

The German physicist and rocket pioneer Wernher von Braun (1912–77) and his team of scientists and engineers designed and built types of military rockets, known as V-1 and V-2 (from the German word *Vergeltungswaffe*, meaning vengeance weapon), to bomb targets in England during World War II. These rockets used a combination of ethyl alcohol and liquid oxygen, along with a few other substances. The V2 achieved a top speed of about 3,500 miles/hour (5,600 km/hr).

After the war, von Braun came to the United States and helped the Americans develop rocket technology. These researchers worked on problems such as the best fuel mixture and how to stabilize a rocket traveling at high speeds. Stability is not a simple issue—a little wobble at high speed can become much worse, resulting in a somersaulting vehicle that is completely out of control. Weight distribution and the geometry of the vehicle are critical.

The result of the work of von Braun and his colleagues was the Saturn V (pronounced Saturn Five—the *V* is the Roman numeral for five, denoting the five powerful engines of the rocket). Used for the Apollo lunar missions, this is the most powerful rocket launched to date. Saturn V had multiple stages, which

The first flight of a Saturn V rocket was the unmanned *Apollo 4* mission, seen here during launch on November 9, 1967. *(NASA/Kennedy Space Center)*

burned through their fuel and then dropped off, relieving the vehicle of their weight. Liquid fuel was used, either a combination of kerosene and liquid oxygen or of liquid hydrogen and liquid oxygen. The rocket attained escape velocity, propelling Apollo astronauts to the Moon.

Liquid oxygen or liquid hydrogen must be cooled in order to condense as liquids, but this is essential to reduce their volume; building a large, massive container would require the rocket engine to accelerate more mass. Mass must be minimized because of Newton's second law, which relates the force, F, acting on an object to its mass, m, and acceleration, a (the rate at which velocity changes), by the equation $F = ma$. Dividing the equation by m, it follows that an object's acceleration equals F/m—an increase in force results in greater acceleration, but the greater the object's mass, the less it will accelerate for a given force. Rockets are as lightweight as possible, and most of the mass is fuel.

Some rockets use solid propellant, often a combination of aluminum powder and ammonium perchlorate (a compound containing the elements nitrogen, hydrogen, chlorine, and oxygen). Solid propellant is simpler and easier to store. Liquid fuels, however, are generally cheaper and easier to control, or throttle, in the engine. The space shuttle uses both solid and liquid propellant—two solid rocket boosters burn solid propellant for acceleration during liftoff, and an external tank containing a liquid oxygen/hydrogen mixture provides fuel for the shuttle's main engines. These powerful rockets lift the nearly 4,500,000-pound (2,000,000-kg) vehicle into space. Exhaust velocity of one of the main engines is about 14,750 feet/second (4,500 m/s)—about 10,000 miles/hour (16,000 km/hr).

Once out in space, there is little resistance to motion—no atmosphere to plow through. As a result, spaceships do not lose much speed. Chemical rockets burn their fuel quickly to reach their desired speed, then they coast. This means that unmanned probes that have been launched to study the outer planets will proceed onward, long after their mission is finished. NASA announced in a May 24, 2005, news release that *Voyager 1,* a probe launched on September 5, 1977, to study Jupiter and Saturn, had neared the limits of the solar system, 8,700,000,000 miles (13,920,000,000 km) from the Sun. From there, it will continue farther into space, crossing vast distances until a collision destroys it or a star or planet "captures" it in a strong gravitational field. The NASA scientist Edward Stone said, "Voyager has entered the final lap on its race to the edge of interstellar space, as it begins exploring the solar system's final frontier."

TAKING THE SLOW BOAT

Voyager 1 took decades to reach interstellar space, and many more decades will pass before it can reach another star. But unmanned probes have the leisure of time—taking the slow boat does not matter.

Humans who venture into interstellar space will want faster transportation. But the universe has a speed limit according to the laws of physics, which numerous experiments have supported. In 1905, the German-American physicist Albert Einstein (1879–1955) postulated that the speed of light is constant—it does not depend on the observer's frame of reference—and he derived some interesting consequences of this and other postulates, as discussed in the sidebar on page 68. One of the results of Einstein's ideas is that no object possessing mass can be accelerated to the speed of light in a vacuum, which is about 186,000 miles/second (300,000 km/s), denoted c. An object can attain a speed of 99.9999 percent of this value—and one can add as many 9s after the decimal point as desired—but it can never reach 100 percent.

Experiments with particle accelerators, which use electric and magnetic fields to boost particles such as electrons and protons to tremendous velocities, show that regardless of how much energy gets pumped into the accelerator, the particles fail to reach light speed. The extra energy does not increase the particle's velocity as much as it increases its apparent mass. According to Einstein's theory, the mass of an object gets closer to infinity as its speed gets close to c. This is why objects can never attain c (or exceed it).

A ship traveling at close to the speed of light will require almost four and a half years to reach the nearest star system, Alpha Centauri, and a lot longer to travel to other stars. Unless researchers find some trick or special condition that circumvents this rule, people will have to live with c as the speed limit.

But even with this speed limit, humans can still reach the stars by one of several methods. One possibility was mentioned in chapter 2—determined colonists can cross the gulf of interstellar space in generation ships (also known as interstellar arks). These large ships would be habitats in which people live out their lives and families would raise their children. Since the journey requires far longer than a human lifetime, generations would come and go before the ship reaches its destination. Engineers might design the habitat to mimic conditions on Earth as much as possible, and the ship's rotation would provide artificial gravity. A generation ship would be a village or city flying through space.

Generation ships would be expensive to build, though, and passengers might not find the prospects of living out their lives in an artificial environment to be too appealing. Another possible way for humans to reach the stars is suspended animation—the reduction or temporary

Jet Propulsion Laboratory (JPL)

In the 1920s and 1930s, rockets enlivened the pages of science fiction magazines, but few scientists other than Robert Goddard considered them anything but fantasies. In the fall of 1936, a group of young people in Pasadena, California, decided to conduct an experiment. Frank Malina, a student at nearby California Institute of Technology (Caltech), and his friends tested a small rocket motor. Despite problems with the oxygen system, they eventually got the motor to work. Encouraged by Caltech professor Theodore von Karman, the young "rocketeers" built more rockets. When their experiments became a little too hazardous for the university's campus, they moved to a new facility in 1940 in the foothills of Pasadena. In 1943, the facility became the Jet Propulsion Laboratory, with von Karman as the first director.

JPL has made many contributions to space exploration. Workers at JPL built and designed *Explorer 1*, the first successful American satellite, launched on January 31, 1958. JPL also played a critical role in the Ranger probes that studied the Moon in the 1960s, the Mariner probes that flew by Venus and Mars in the 1960s and 1970s, the Viking probes that landed on Mars in 1976, the Voyager probes that reached the outer planets in the 1980s, the *Mars Pathfinder* that roamed Mars in 1997, and many others.

Today, JPL is a government-funded laboratory, managed by Caltech for NASA. Current and future projects in which JPL is participating includes *Kepler,* a satellite to search for Earth-like extrasolar planets (see chapter 1), and the *James Webb Space Telescope,* which is expected to launch in 2013. Research projects include deep space navigation, robotics, communication networks to keep track of distant spacecraft, and many others.

suspension of an organism's physiological functions. Certain animals escape harsh conditions and times of scarcity by becoming dormant, and mammals such as bears survive the winter by hibernating. Perhaps humans can greatly extend their lifetimes with a similar technique; cold storage or freezing, for example, may preserve the body and brain. A few people have even had vital tissues preserved after death in the hope that future scientists will revive them. This practice is called cryonics.

The problem with freezing is that ice damages cells and tissues. Although some animals such as a few species of frogs have evolved physiological mechanisms to protect themselves from this kind of damage, humans do not have this capacity. Suspended animation for space travel does not look promising, although future advances in biology may one day enable it.

Another possible solution to the problem of lengthy journeys is more feasible. This solution involves the physics of traveling at speeds a substantial fraction of the speed of light; this will be discussed in the following section.

To minimize the time required for the journey, interstellar explorers need the fastest possible ships. Current rocket technology does not come close to approaching c. Saturn V triumphantly attained Earth's escape velocity, but this speed is only about 0.004 percent of the speed of light. There is much room for improvement.

At the forefront of efforts to build improved propulsion devices is the Jet Propulsion Laboratory (JPL) in Pasadena, California. This NASA laboratory designs and builds satellites and space probes, manages instrument systems and collects and analyzes data, and studies engines and advanced propulsion concepts. The sidebar offers more information on this laboratory.

JPL has recently tested a propulsion system called an ion drive. Ions are charged particles. Instead of creating hot, expanding gases with chemical reactions, an ion propulsion system ionizes a gas such as the element xenon, creating charged atoms that respond to electric fields. The motor uses electric fields to accelerate the ions, which are emitted as exhaust and propel the spacecraft forward by Newton's third law. An ion drive powered the probe *Deep Space 1,* launched on October 24, 1998, which performed well.

Ion propulsion systems in operation today cannot achieve rapid acceleration, but they are extremely efficient and operate over longer periods of time. These systems provide a series of gentle nudges rather than the burst of a chemical rocket, the propellants of which are rapidly depleted. If ion drives make use of solar power, they can keep running as long as they are near enough to the Sun, or another star, to receive sufficient energy to ionize and accelerate the gas. The European Space Agency (ESA) has built an ion engine called SMART-1, which uses solar-powered ion propulsion. As ESA notes in an article, "The Magic of Ion Engines," posted on their Web site, "An ion engine can go on pushing gently for months or even years—for as long as the Sun shines and the small supply of propellant lasts." Slow and steady may win the race.

MAKING TIME GO SLOWER

If researchers manage to design engines that reach extremely fast speeds, either by a burst of rapid acceleration or a longer period of smaller increments in speed, then human explorers may be able to travel to nearby stars. And if the ship's speed is an appreciable fraction of c, the journey will not seem to take very long at all. This effect is called time dilation—an observer traveling at high speed experiences a slowing of time when compared to the clocks of a stationary observer.

Slowing of time does not refer to clocks being set at a different time, but involves clocks running at different rates. Different clock settings are familiar concepts, since Earth is separated into local time zones; for example, the clock of a person living in Atlanta, Georgia, runs exactly three hours ahead of the clock of a resident of Seattle, Washington. These zones coordinate time settings so that everyone can agree what time it is, although the hours are shifted to keep day and night at the appropriate local time. Although the hours are shifted, the clocks run at the same rate.

Time dilation appears counterintuitive. People are accustomed to having the same time frame as everyone else, so that, for example, the exact instant that a particular event occurs is the same for all observers. If lightning strikes a certain spot, a person unfamiliar with physics might think that all observers will agree on the time at which the lightning struck. But this is not true. Two observers who are in relative motion at constant velocity—in other words, one observer is moving at a constant speed with respect to the other—will disagree.

The physicist who proposed this odd theory was Einstein. In 1905, Einstein published his ideas on invariance in physics. Einstein believed that the laws of physics should be the same, or invariant, for all observers; no matter where an observer is standing, or if the observer is moving or not, the laws and equations of physics should be the same for everyone. In order to formulate the laws and equations to be invariant, Einstein postulated that the speed of light was constant and the same for all observers. But he also found that certain measurements would have slightly different values for observers who are in relative motion. One of these measurements is simultaneity. Two events that occur simultaneously to one observer at rest might occur at different times to another observer who is in motion, and vice versa. Einstein's theory is known as the special theory of relativity, as described in the following sidebar.

According to the special theory of relativity, clocks run slower at high speed, as judged by the rate of stationary clocks. The special theory of relativity applies only under special conditions—the speed is constant, meaning there is no acceleration. (In the attempt to generalize these ideas and incorporate acceleration, Einstein wound up with another theory, the general theory of relativity. This theory involves acceleration and gravity, and Einstein discovered that clocks run slower in a gravitational field as well.)

Suppose Albert the astronaut blasted off into space on a high-speed rocket and left his twin brother, Holmes, on Earth. The special theory of relativity says that Albert's clocks will run more slowly than Holmes's—including biological clocks. This means that Albert will age less rapidly than his twin. Although Albert's trip may take 20 years according to Earth-bound clocks, his own clocks may read 10 years. When Albert returns, he will be 10 years younger than his twin.

This effect is sometimes called the twin paradox, although it is not a paradox—there is no contradiction. Experiments in particle physics as well as clocks traveling on high-speed aircraft support the theory, so Albert's clock will run more slowly.

A paradox does arise, however, when one considers that only relative motion is important in the special theory of relativity. In other words, it does not matter which observer is moving, the only important requirement is that one observer is in motion relative to the other. To an observer, P, standing beside a railroad track, a passing train is moving past at 40 miles/hour (64 km/hr). But to an observer, P', sitting on the train, P seems to be flying past at 40 miles/hour (64 km/hr). Because of the invariance of the laws and equations of physics, the special theory of relativity recognizes no difference in the two situations. Each of the

Special Theory of Relativity

Einstein was only 26 years old when he published the special theory of relativity in 1905. The young Einstein had a penchant for thinking in original and creative ways, a style known today as "thinking outside the box." Suppose that an observer P is standing beside a railroad track, and a train comes along moving at velocity v and carrying observer P'. The figure illustrates this situation. Imagine that lightning strikes the front (B) and rear (A) of the train just as the train passes the stationary observer. In order to see the lightning bolts, the light from the event must travel to the observer's eyes (or camera). If the bolts struck at equal distances from observer P, they will travel this distance in the same amount of time. Suppose P sees the bolts at the same time. To this observer, the two bolts occurred simultaneously.

But Einstein realized that observer P' will see something different. Because P' is moving, this observer travels some distance toward B, the point at which one bolt struck, and away from A, the point at which the other bolt struck. The speed of light, as Einstein postulated, is the same for all observers—even if an observer is moving toward or away from a light source, the speed of light is the same. This means that the light from the bolt that struck B will arrive slightly before the other bolt, as determined by observer P'. The two observers disagree on when the lightning bolts struck!

Einstein discovered equations that determine the time difference between observers in relative motion. He found that from the perspective of stationary observers, time runs slower for observers in motion; in other words, clocks seem to run more slowly when they are in motion. This is time

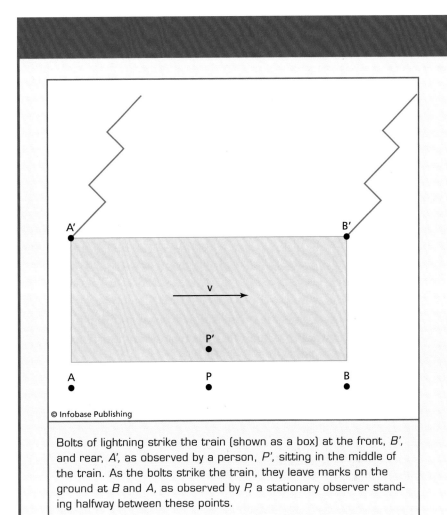

Bolts of lightning strike the train (shown as a box) at the front, *B'*, and rear, *A'*, as observed by a person, *P'*, sitting in the middle of the train. As the bolts strike the train, they leave marks on the ground at *B* and *A*, as observed by *P*, a stationary observer standing halfway between these points.

© Infobase Publishing

dilation. Many people are skeptical when they first hear this theory, but plenty of evidence supports it. For instance, the average lifetime of a quickly moving particle extends well beyond a stationary one. In addition, highly accurate atomic clocks placed on jet airplanes run slightly more slowly than their counterparts sitting in a laboratory.

observers is right, and neither one can do a physics experiment to prove that he or she is in motion and the other is at rest. Motion is relative, not absolute.

The twin paradox arises because Holmes can claim that he was the one that was moving, so his clocks should be slower. This would mean that Holmes is the younger twin. And that is a paradox—both cannot be younger than the other.

Resolving the paradox is complicated, but it does not falsify the special theory of relativity. The special theory applies to constant speed, and Albert, the space-faring twin, clearly accelerated, both in speed and direction (when he turned around to come home). Acceleration breaks the symmetry, and its effects are easily noticed by observers—an observer is not able to distinguish whether he or she is at rest or moving at a constant rate of speed, but an observer can detect acceleration from the push in the back of the seat experienced when a car spurts forward, for example. The twin "paradox" is not a contradiction because acceleration distinguishes one of the two experiences. According to Einstein's theories, Albert will be younger than his twin when he arrives home.

STARSHIP PROPULSION

Time dilation is a real effect, but only becomes significant at exceptionally high speeds. Clocks on the *International Space Station,* which orbits at a speed of about 17,500 miles/hour (28,000 km/hr), run slower due to the special theory of relativity by a factor of about 1.0000000003, which is negligible. At 18,600 miles/second (30,000 km/s)—10 percent of c—the factor is 1.005, which reduces a week to the equivalent of 6 days, 23 hours, and 10 minutes. For a speed of 99 percent of c, the factor is 7.09—a week on Earth amounts to only about a day on board the ship. Since all processes on board the ship would be slowed, passengers would not notice any effects. To a passenger's perception, everything would seem normal.

Humans can reach the stars—even ones at vast distances—if their ship can achieve a velocity that is close to the speed of light. Time dilation "extends" the life span, at least in terms of Earth's calendar. But society may have changed drastically by the time the explorers get back.

Engineers have yet to build a propulsion device that comes close to attaining the required speeds, and researchers do not know what kind

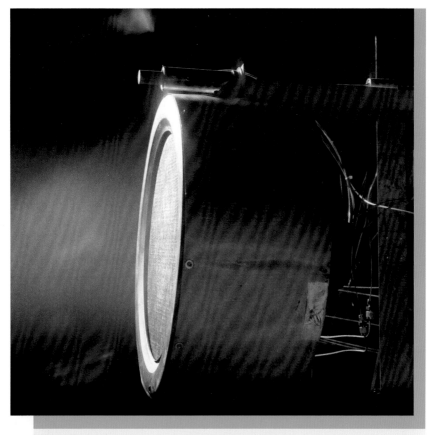

This photograph of an ion engine shows a bluish glow coming from the emitted ions. *(Jet Propulsion Laboratory)*

of machine would offer the best performance. Several engines are being studied, including ion engines and propulsion using nuclear energy.

Ion engines, with their efficiency as well as slow and steady acceleration, have several advantages. Attaining high speeds must be accomplished gradually, otherwise the strong forces associated with rapid accelerations would crush the passengers. Acceleration magnitudes are often measured in units of G, the acceleration experienced by an object on the surface of Earth. One G is 32 feet/second² (9.8 m/s²)—the velocity of a dropped object increases at a rate of 32 feet/second (9.8 m/s) for every second it falls (in the absence of other forces such as air resistance). This equals the familiar tug of gravity that everyone feels. An acceleration of 2 G is twice as high, and as a result, a person would feel twice as heavy.

Another one of Einstein's discoveries is that the effects of acceleration—the push in the back as a car increases speed, for example—are indistinguishable from gravity. This means that a ship accelerating at 1 G would seem to have "normal" gravity in which dropped objects would fall toward the floor (or, more precisely, toward the rear of the ship) at a rate of 32 feet/second^2 (9.8 m/s^2). Slightly higher accelerations are endurable, but uncomfortable. Space shuttle astronauts, for example, experience brief episodes of about 3 or 4 G. People lose consciousness at around 12 G, and even higher accelerations can crush and kill.

A steady, gentle acceleration of 1 G would be ideal from a passenger's point of view. The ship would take awhile to get up to speed, but it would also provide a comfortable environment in the process.

One of the difficulties with this scenario is fuel. Long-lasting accelerations use an excessive amount of fuel, which must be carried at launch time unless fuel can be replenished in-flight. A large mass of fuel makes launches prohibitively expensive as well as dangerous. Carrying a great deal of fuel during the journey also poses serious risks of fire and explosion. In space, even minor mishaps can be fatal.

To minimize the mass while providing the power for prolonged accelerations, the fuel should be rich in energy though not bulky. Several options are possible. The best option would allow conversion of every bit of the fuel into energy; this option is possible under special circumstances, as discussed in the section below titled "Engines That Use Antimatter." Another option that is currently feasible is energy from an atom's nucleus.

Nuclear reactions such as fission, in which a nucleus splits, and fusion, in which nuclei join, yield a huge amount of energy. Fission is the simpler of the two reactions to control and has been used in nuclear reactors to generate electricity for consumers as well as propelling certain ships and submarines. Space agencies have already incorporated nuclear power into the energy supply of probes that must travel lengthy distances from the Sun, such as probes that visit the outer planets. (Probes that do not stray far from the Sun tend to use solar energy since they can capture enough of the Sun's energy to function.) But this use of nuclear power provides power for scientific instruments and other electrical apparatus rather than propelling the vehicle through space.

NASA recently began a program called Project Prometheus aimed at exploring nuclear propulsion. (In Greek mythology, Prometheus was a Titan who stole fire from the gods and gave it to people.) On April 18, 2005, the agency issued a news release asking for public comment on the project. The news release stated that they were "evaluating the possibility of developing a space nuclear reactor to supply future exploration spacecraft with a significant increase in onboard power and spacecraft propulsion capability. Such an increase in power would enable missions to the outer reaches of the solar system and beyond as well as substantially increasing the amount of science per mission." But due to budget constraints, competing projects, and opposition to nuclear power in general, NASA has not actively pursued this program. One of the main concerns is that an accident during liftoff of a nuclear-powered spacecraft could result in the release of hazardous material, similar to the 1986 explosion of the Chernobyl nuclear reactor in Ukraine. Another problem for manned spacecraft is that shields must be installed to protect the crew from the reactor, adding to the bulk of the vehicle.

Regardless of the source of propulsion, a major concern with high-speed travel is the threat of collisions. Although interstellar space is nearly a vacuum, it does contain numerous atoms, molecules, and larger bits of debris. If a ship hits a rock while cruising at 93,000 miles/second (150,000 km/s)—$0.5c$—the effect is equivalent to a projectile striking the vessel at $0.5c$. Such a collision would cause catastrophic damage even if the rock is tiny; a pebble, for example, could pass all the way through the ship, rupturing the fuselage and allowing air to escape. The upside is that ships may be able to scoop up interstellar material as fuel, especially the relatively abundant hydrogen atoms, but future scientists and engineers will have to concoct a mechanism to deflect larger and potentially deadly objects.

SAILING WITH LIGHT AND THE SOLAR WIND

The problem of the need for a tremendous amount of fuel for interstellar travel can be avoided if the ship somehow procures the required energy as it goes along. One method of accomplishing this is to scoop up interstellar material, but another possibility is to use solar energy in a unique manner.

Electromagnetic radiation from the Sun has both wave and particle aspects. The particles, called photons, behave as little packets of energy that either bounce off or get absorbed when they encounter objects. Photons therefore exert a pressure. One photon does not have much energy and creates only a tiny pressure, but the effect adds up when photons are abundant.

An additional source of pressure is the *solar wind*. The outer layers of the Sun eject a steady stream of particles, mostly protons and electrons, because the high temperature propels them out of the Sun's gravitational field. These particles continue away from the Sun at high speeds, shooting past Earth at about 310 miles/second (500 km/s).

When acting over a large area, photons and the solar wind can have noticeable effects. The tail of a comet, for example, points away from the Sun due to this pressure. Some creative researchers have noted that this pressure might be harnessed to drive a spaceship. The idea is to attach a gigantic sail—a solar sail—to catch the solar wind and radiation. Astronauts would sail through space similar to the old sailing vessels that plied the seas. The sail must be expansive in order to get enough "wind," and should be highly reflective—polished like a mirror—so that light reflects from the surface rather than getting absorbed. (Reflection transfers more momentum to the sail.)

A thin sail, even if it covers a huge area, can be made from lightweight material, which reduces the mass at liftoff. Taking off in Earth's thick atmosphere with such a sail would be unwise, but the sail can be "stitched" together in space, or perhaps rolled up and fitted inside a rocket, and unfurled in space. Solar sailing would not generate much acceleration, but it would be steady. Collisions with debris and other objects would probably occur if the sail covers a wide area, but holes in the sail would not be catastrophic and could be repaired as needed.

An early version of a spaceship operating with a solar sail was *Cosmos 1*, a joint project of the Planetary Society, an organization devoted to space exploration, and Cosmos Studios, a company led by Ann Druyan, author and widow of the late astronomer Carl Sagan (1934–96). Launched on June 21, 2005, attached to a submarine-launched rocket, the ship was to spread its sails in space. The Planetary Society described the goal of the project on their "Solar Sailing" Web page: "*Cosmos 1* was not intended to go to the stars, but only to prove that solar sailing was possible. Once the spacecraft had entered Earth orbit, it was to

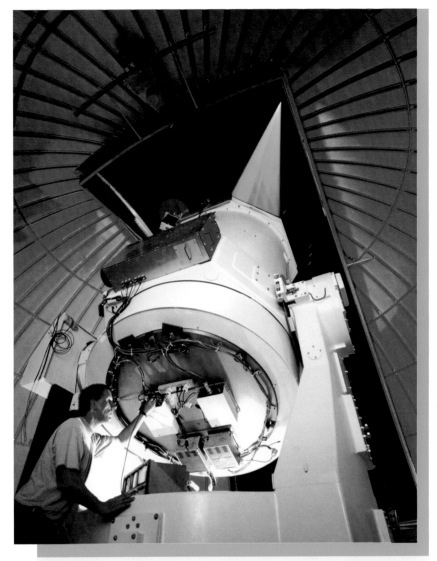

This laser is used to create an artificial "star" that helps astronomers calibrate their telescopes. A similar but more powerful laser could possibly drive a spacecraft long distances. *(D. Parker/Photo Researchers, Inc.)*

raise its altitude through solar sailing. Any measurable increase in the spacecraft's orbit would have been considered a success." But the rocket failed, and *Cosmos 1* was lost.

Solar sailing might be enough to propel a spaceship through the solar system, but what about interstellar space? The great distance would eventually render the pressure too small to be of much use. In 1989, the physicist and science fiction author Robert L. Forward (1932–2002) described the use of a laser beam as the means of propulsion. The concept is the same as a solar sail, except the radiation pressure comes from an intense laser rather than the Sun or another star.

The term *laser* stands for light amplification by stimulated emission of radiation. When pumped with energy, photons in the laser stimulate the production of other photons of the same wavelength, which are emitted in a coherent beam. Laser beams do not spread nearly as much as light from sources such as a flashlight, so the energy remains concentrated in a small space over long distances. This is ideal for a spaceship's sail.

Scientists on Earth have already bounced laser beams off the Moon in order to gauge the distance between Earth and the Moon with great precision. (The laser beams bounced off special reflectors placed on the Moon's surface by Apollo astronauts.) In practice, laser propulsion is a valid idea. The laser could be set up in orbit around Earth perhaps and train its beam on the spaceship. But the power to drive a spaceship interstellar distances would be enormous and beyond the capacity of current devices.

ENGINES THAT USE ANTIMATTER

Solar sails, laser beams, and the use of interstellar hydrogen may be able to provide the slow, steady acceleration needed to reach the stars, but none of these techniques may be wholly sufficient. Some other kind of engine might be necessary to supplement the main drive. If this engine and its fuel are highly efficient, the other methods might not be needed after all. Engines that use antimatter are prime candidates.

NASA is considering development of antimatter engines for a manned mission to Mars, and if the technique works, perhaps further journeys would be feasible. In a press release issued on April 14, 2006, NASA noted, "Most self-respecting starships in science fiction stories use antimatter as fuel for a good reason—it's the most potent fuel known."

Antimatter is similar to matter except certain properties such as electric charge are reversed. All particles have an antimatter counterpart, known as its antiparticle. (Some particles such as the photon are their own antiparticle.) For example, the positron has the same mass and

Antimatter

In 1928, the British physicist Paul Dirac (1902–84) formulated a set of equations that had a strange result—they predicted the existence of a particle with the same mass as an electron but the opposite charge. Four years later, the American physicist Carl Anderson (1905–91) found the track of a positively charged electron, or positron, in a particle detector called a cloud chamber. This positron was first piece of antimatter discovered.

Antiparticles behave in a similar manner as ordinary particles, except that certain features such as electric charge are reversed. Some researchers describe antimatter as the "mirror" of matter, since it seems to be the equivalent of matter except for the reversed features, as in a mirror image. All the matter that scientists have examined in the universe is made of ordinary matter instead of antimatter. One reason for this is that matter and antimatter disappear when they meet, replaced in general by two photons moving at opposite directions. Physicists hypothesize that the universe is made almost exclusively of matter because there was a slight preponderance of ordinary matter over antimatter at the time of creation; after most of the particles and antiparticles annihilated one another, a slight excess of matter remained. But no one knows why more matter than antimatter was created.

Physicists produce and study antiparticles in high-speed collisions that occur in particle accelerators (also known as "atom smashers"). The debris of the collisions sometimes includes antiparticles. This process of creating antimatter is the opposite of matter-antimatter annihilation. When a particle and its antiparticle meet, they vanish in a burst of energy, but on some occasions a tight concentration of energy—which may occur in violent events such as high-speed collisions—gets transformed into a particle and its antiparticle. This is possible because of the relation discovered by the German-American physicist Albert Einstein (1879–1955), between energy, E, mass, m, and the speed of light in a vacuum, c: $E = mc^2$. The process of generating particles and antiparticles is called pair production.

Artist's conception of an antimatter-powered spaceship, with a circular particle accelerator to generate the fuel *(Christian Darkin/Photo Researchers, Inc.)*

charge magnitude as an electron, but the sign of the charge is reversed—positrons are positively charged instead negatively charged, as is the electron. (The term *positron* is short for positive electron.) When matter and antimatter meet, they are annihilated in a burst of energy. The sidebar on page 77 provides more information on this exotic substance.

Matter-antimatter annihilation is such an efficient source of energy because the mass of both particles gets transformed into high-energy photons. This is complete conversion; 100 percent of the fuel is used, unlike other methods to generate power, all of which leave a substantial residue that is wasted. Nuclear power, for example, converts only about 1–3 percent of the mass of the fuel into energy.

But there are problems with a potential matter-antimatter engine. High-energy photons are dangerous, so crew members must be shielded. Storing antimatter also poses difficulty since any contact with matter results in annihilation. Antimatter must be confined with electric or magnetic fields rather than conventional containers. There is also the issue of cost. Antimatter production occurs only under high-energy experiments such as the collisions of particle accelerators, which are extremely expensive to operate.

NASA scientists are working on the best options and a novel design. According to the April 14, 2006, news release, "Previous antimatter-powered spaceship designs employed antiprotons, which produce high-energy gamma rays when they annihilate. The new design will use positrons, which make gamma rays with about 400 times less energy." Gerald Smith, a researcher at Positronics Research, a company based in Santa Fe, New Mexico, spoke about the cost. "A rough estimate to produce the 10 milligrams of positrons needed for a human Mars mission is about 250 million dollars using technology that is currently under development." Although the price will fall as researchers learn how to produce antimatter more efficiently, a trip to the stars would cost much more.

WORMHOLES—TUNNELS IN SPACE AND TIME

The forbidding distances separating the stars tend to dampen any optimistic scenarios of interstellar travel. Even the most enthusiastic supporter of space exploration must acknowledge the difficulties. But there is a slight chance that the vast distances may prove irrelevant if people can find shortcuts through space. One possibility is a *wormhole.*

Wormholes are "tunnels" connecting different parts of space and time. In his general theory of relativity, Einstein conceived of gravitation as a warping of space and time. Einstein's picture of the universe consists of four dimensions—three of space and one of time—known as *space-time,* in which the presence of a massive body curved or warped the space-time in its vicinity. This curvature, Einstein proposed, was the reason why objects with mass attract one another; instead of considering gravitation as a force acting at a distance, Einstein visualized it geometrically.

Much evidence has been found to support the general theory of relativity. One of the most astonishing of its predictions was that the universe is expanding. Even the theory's discoverer had trouble believing this prediction, so Einstein inserted a term in the complicated

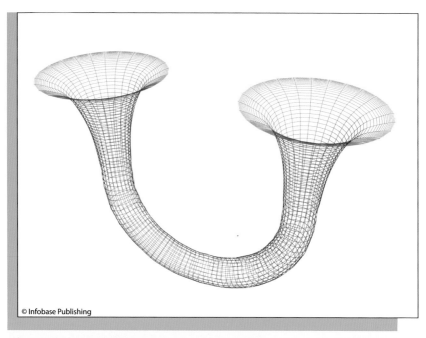

© Infobase Publishing

A wormhole connects two regions of space-time.

equations so that they instead represented a stationary universe. (See the sidebar on page 155.) But astronomers discovered otherwise, and Einstein admitted his mistake. The general theory of relativity is one of the foundations of modern physics and is used as a large-scale model of the universe.

A number of interesting objects appear in the solutions to the equations of general relativity. One object is a black hole—an object with a gravitational field so strong that nothing, not even light, can escape it. Another object is a wormhole, which forms a sort of tunnel through space-time. As illustrated in the figure, this tunnel might be a shortcut connecting one point of space-time to another.

But even if wormholes exist, no one knows if they would be traversable. The equations indicate that these objects would be highly unstable. If pieces of matter such as a dust cloud, a rock, or an astronaut tried to enter, the wormhole may snap shut, with unfortunate consequences.

In a 1988 paper, the California Institute of Technology physicists Michael Morris, Kip Thorne, and Ulvi Yurtsever described a theoretical method of propping a wormhole open. These researchers found that the addition of significant masses and electric fields could keep the "tunnel" clear, though some of these additions must have strange properties, such as exerting negative pressure—a vacuum. But other kinds of wormholes might be easier to traverse without the need for unusual forms of matter to shore them up.

Wormholes represent valid solutions to the complex equations of general relativity, but this does not imply that they exist—the theory permits their existence but does not require it. Astronomers have found no evidence of a wormhole, although this does not prove they do not exist. Yet if wormholes exist, some of their properties suggest troubling paradoxes. For instance, wormholes are not only shortcuts in space but also in time, so a wormhole, if one exists, may act as a time machine. Time travel is exciting but invokes paradoxes such as the possibility of meeting one's self, or accidentally creating a situation in which one's parents fail to reproduce. These conundrums seem impossible, so time travel would also appear to be impossible. If so, then perhaps wormholes have no basis in reality.

CONCLUSION

People have explored only a small corner of the universe, which resides in the vicinity of a small star with eight planets. Astronomers and physicists have learned a lot with their telescopes and their imaginative theories, such as Einstein's general theory of relativity, but there is much left to be discovered. No one knows what awaits future interstellar explorers as they roam the vast distances between the stars. They may find many strange objects, possibly even more unusual than wormholes.

Shortcuts through space and faster-than-light travel often form the backbone of fascinating science fiction stories. These devices make it possible for authors to use plots involving galactic empires and warring factions, but most of these devices violate the laws of physics or permit contradictions that are extremely unlikely to exist in nature. There is no reason to assume that currently accepted scientific theories will stand forever, but the casual dismissal of the laws of physics as they are now understood strains belief.

Other advanced concepts of interstellar exploration tend to follow the laws of physics more closely. Even with this constraint, many exciting developments are imaginable. One concept is the faster-than-light motion of hypothetical particles called tachyons. (The term *tachyon* derives from the Greek word *tachys,* meaning "swift.")

According to Einstein's theories, no object with mass can ever be accelerated up to or beyond c, the speed of light in a vacuum. Only photons, which have no rest mass, can reach this speed. As a particle or object with mass approaches c, its mass seems to increase, making subsequent acceleration even more difficult. At c, the mass would become infinite, which is impossible. Therefore no object with mass can reach a speed equal to c.

But there is another possibility. Suppose a particle has always been traveling beyond c. No one could raise the objection of an impossible increase of mass with acceleration because a tachyon was not accelerated up to or beyond c—it was "born" exceeding this velocity. Tachyons could presumably never slow down to c, since this would entail infinite mass. The speed of light would be the minimum rather than the maximum for tachyon speed.

No one has ever observed a tachyon. Proof of these particles, if they exist, might require a new and entirely different set of instruments, otherwise the faster-than-light objects may remain invisible. Tachyons might be useful in communication and other transmission devices, but it is not obvious how they might be employed. To make use of an object, some sort of device or instrument must interact with it in some way. The human eye, for example, absorbs light, which evokes a chain of activities in the brain that creates visual perception. Some kind of detector or transmitter would be essential in any tachyon device. But no one is certain how an instrument could interact with a tachyon without at least briefly bringing its speed below the speed of light. And for a tachyon, no speed below c is permitted.

Another advanced concept is teleportation. Teleportation is the transportation or transfer of matter across space with no apparent medium. On the television series *Star Trek,* people and inanimate objects are often "beamed" from a starship to a planet and back again with special teleportation devices. The object to be beamed disappears at one site and reappears in the desired location. Although this process seems highly improbable, a similar technique known as quantum teleportation is within the realm of the laws of physics.

Quantum teleportation uses principles of *quantum mechanics,* such as the ability of an object to exist in multiple states at the same time, to make replicas of the state of small objects such as photons or atoms. The particles seem to jump across space. For example, FoxNews.com reported on January 26, 2009, that the University of Maryland researcher Christopher Monroe and his colleagues succeeded in teleporting a quantum state from one atom to another across a distance of about 3.3 feet (1 m). Such feats require incredible skill and have only been successful for tiny particles and short distances.

Advanced ideas such as teleportation, laser sails, and antimatter engines are only in their early stages of development. As with all new ideas, they may or may not work. But scientists at the frontiers of space and astronomy are continuing to investigate options to permit future generations of explorers to reach the stars. Achieving a practical means of interstellar travel would greatly enhance the current understanding of the universe and the place of humanity within it.

CHRONOLOGY

1687	The British physicist Sir Isaac Newton (1642–1727) publishes his discoveries on the laws of motion.
1903	The Russian teacher and scientist Konstantin Tsiolkovsky (1857–1935) publishes the basic science of rocket flight.
1905	The German-American physicist Albert Einstein (1879–1955) publishes the special theory of relativity.
1926	The American physicist and inventor Robert H. Goddard (1882–1945) launches the first liquid-fueled rocket.
1932	The American physicist Carl Anderson (1905–91) discovers antimatter.
1936	The initial propulsion experiments are conducted at what would become the Jet Propulsion Laboratory.

1958	Project Orion, led by the physicist Theodore Taylor, becomes the first program to study nuclear propulsion for spacecraft.
	The U.S. government establishes NASA.
1960	NASA Lewis Research Center (since renamed as the Glenn Research Center) develops the first ion drives.
1967	The first launch of a Saturn V rocket (unmanned) occurs at Kennedy Space Center in Florida.
1989	The physicist and science fiction author Robert L. Forward (1932–2002) proposes an advanced propulsion concept based on a laser sail.
1998	The spacecraft *Deep Space 1* successfully tests an ion engine.
2005	The probe *Voyager 1* reaches the edge of the solar system.
	Cosmos 1, a space vehicle designed to test the concept of solar sails, is destroyed when the rocket carrying it into space fails to reach orbit.
2006	NASA explores the possibility of an antimatter engine to propel a manned vessel to Mars.
2009	The University of Maryland researcher Christopher Monroe and his colleagues teleport a quantum state from one atom to another across a distance of about 3.3 feet (1 m).

FURTHER RESOURCES

Print and Internet

Bonsor, Kevin. "How Antimatter Spacecraft Will Work." Available online. URL: http://science.howstuffworks.com/antimatter2.htm. Accessed July 27, 2009. An entry at the How Stuff Works Web site, this article discusses the possibility of antimatter engines.

European Space Agency. "Magic of Ion Engines." Available online. URL: http://www.esa.int/SPECIALS/SMART-1/SEMLB6XO4HD_ 0.html. Accessed July 27, 2009. ESA describes the ion engine called SMART-1, which stands for the project called Small Missions for Advanced Research in Technology.

FoxNews.com. "Scientists Teleport Matter More Than Three Feet." January 26, 2009. Available online. URL: http://www.foxnews.com/ story/0,2933,482264,00.html. Accessed July 27, 2009. The University of Maryland researcher Christopher Monroe and his colleagues have teleported a quantum state from one atom to another across a distance of about 3.3 feet (1 m).

Gilster, Paul. *Centauri Dreams: Imagining and Planning Interstellar Exploration.* New York: Copernicus Books, 2004. Gilster discusses what it would take for astronauts to launch a successful mission to Alpha Centauri or another nearby star.

Morris, M. S., K. S. Thorne, and U. Yurtsever. "Wormholes, Time Machines, and the Weak Energy Condition." *Physical Review Letters* 61 (1988): 1,446–1,449. The authors discuss wormholes and their potential applications.

National Aeronautics and Space Administration. "A Brief History of Solar Sails." Available online. URL: http://science.nasa.gov/headlines/ y2008/31jul_solarsails.htm. Accessed July 27, 2009. This article describes the development of the concept of sailing in space with the aid of radiation.

———. "NASA Seeks Public Input on Prometheus." News release, April 18, 2005. Available online. URL: http://www.nasa.gov/vision/ universe/solarsystem/prometheus_peis.html. Accessed July 27, 2009. NASA briefly describes the goal of producing a space nuclear reactor and calls for public comment on the project.

———. "New and Improved Antimatter Spaceship for Mars Missions." News release, April 14, 2006. Available online. URL: http://www.nasa. gov/exploration/home/antimatter_spaceship.html. Accessed July 27, 2009. NASA announces a research project aimed at developing antimatter engines with positrons as fuel.

———. "Voyager Enters Solar System's Final Frontier." News release, May 24, 2005. Available online. URL: http://www.nasa.gov/vision/

universe/solarsystem/voyager_agu.html. Accessed July 27, 2009. NASA announces that *Voyager 1* has entered the boundary of the solar system.

Neufeld, Michael. *Von Braun: Dreamer of Space, Engineer of War.* New York: Knopf, 2007. This biography of pioneering rocket scientist Wernher von Braun covers the life, technology, and the sometimes controversial decisions von Braun made over his long career.

Planetary Society. "Solar Sailing." Available online. URL: http://www. planetary.org/programs/projects/solar_sailing/. Accessed July 27, 2009. The Planetary Society describes *Cosmos 1,* a vehicle that would have tested a solar sail but was lost when the rocket carrying it into space failed.

Rogers, Lucy. *It's ONLY Rocket Science: An Introduction in Plain English.* New York: Springer Science and Business Media, 2008. Written by an engineer, this book introduces the intricacies of rocket science in simple terms. The concluding chapter discusses advanced propulsion systems of the future.

Spangenburg, Ray, and Diane Kit Moser. *Wernher von Braun: Rocket Visionary.* Rev ed. New York: Facts On File, 2008. An impressive account of the rocket scientist's contributions to space exploration during the period between the 1930s and the 1970s.

Stern, David P. "From Stargazers to Starships." Available online. URL: http://www-spof.gsfc.nasa.gov/stargaze/Smap.htm. Accessed July 27, 2009. This resource contains a large number of articles on all aspects of space travel.

Web Sites

Beyond Rocketry: Science@NASA. Available online. URL: http://science. nasa.gov/BeyondRocketry.htm. Accessed July 27, 2009. Features of this Web site discuss advanced propulsion concepts such as solar sails.

Glenn Research Center. Available online. URL: http://www.nasa.gov/ centers/glenn/home/. Accessed July 27, 2009. This Web site provides news and information on the latest research of this NASA laboratory.

Jet Propulsion Laboratory (JPL). Available online. URL: http://www.jpl. nasa.gov/. Accessed July 27, 2009. JPL's Web site contains news and information on the laboratory's missions and scientific research.

4

GRAVITATIONAL WAVES

The points of light in the night sky spurred curious observers to develop the science of astronomy. As scientists studied the light of distant stars or the light reflected from other planets in the solar system, their understanding of the universe grew rapidly. Visible light is only one form of electromagnetic waves, also known as electromagnetic radiation—light occupies a small portion of the electromagnetic spectrum. When astronomers developed instruments to detect other parts of the spectrum, such as radio waves, they opened other important windows into the universe. Additional windows would also greatly benefit astronomy. This is one of the reasons why astronomers are interested in a type of wave called a gravitational wave.

Gravitational waves are propagating disturbances set up by certain motions of massive objects. Scientists have not yet proven the existence of these waves because no one has directly observed one. But the theory of general relativity, one of the foundations of modern physics, predicts this phenomenon, and careful measurements support this prediction, although indirectly, by the effects the emission of gravitational waves have on a particular star system.

Direct observation of these small and difficult-to-detect waves would conclusively prove their existence. Instruments capable of observing gravitational waves would also enable researchers to conduct important tests of general relativity, as well as providing a great deal of information on the motion and nature of large objects in the universe. Gravitational waves offer a powerful way to peer into deep space.

This chapter describes gravitational waves and the situations that theoretically produce them. Attempts to observe gravitational waves, as well as the prospects for success, are discussed. The chapter will also cover the ways in which gravitational waves are useful in astronomy.

INTRODUCTION

Waves are propagating disturbances. Consider what happens when someone throws a stone in a pond, for example. At the point where the stone enters the surface, water is briefly displaced. This water jostles its neighbors. Since water is nearly incompressible—it cannot easily be squeezed into a smaller volume—the energy gets transmitted down the line rather than quickly absorbed. Neighbors jostle neighbors, and so on. As a result, the initial disturbance propagates in all directions from the origin, and a wave sweeps across the surface of the pond.

In a wave, it is the disturbance that travels or propagates, not the material in which the waves travel. A ripple in a pond, for example, is a propagating disturbance of the water. The energy of the motion is what moves outward from the origin. Although water molecules participating in the wave move up and down and jostle the neighboring molecules, their contribution is localized; they do not travel along with the wave.

Another example is the wave that propagates in crowded stadiums as fans in consecutive sections stand up and then sit down again. The wave in this case is the up-and-down motion of the fans, none of whom change seats in the process.

The material in which waves propagate is called the medium. For a ripple in a pond, the medium of the wave is water—water "carries" the wave. Fans in the stadium are the carriers of the waves that roll through the crowd.

Important properties of waves include amplitude and wavelength, as illustrated in the figure, which shows a vibrating string. The amplitude of a wave is the maximum displacement. Wavelength is the distance between any two corresponding points in the wave's cycle, such as the distance from one crest to the next. Frequency indicates the number of cycles of the wave per second, usually described in units of hertz, or cycles per second (named for the German physicist Heinrich Hertz [1857–94], who investigated electromagnetic waves). Wave-

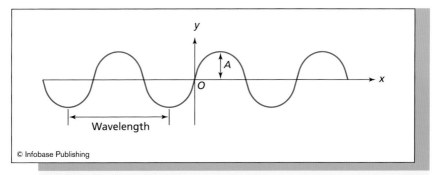

Amplitude, *A*, is the maximum displacement. Wavelength is the length of one cycle, equal to the distance between any two corresponding points in the wave's cycle.

length and frequency are related—the higher the frequency, the longer the wavelength, and vice versa. The product of wavelength and frequency is the speed of the wave. Characteristics of the medium, such as its compressibility and resistance to motion, generally determine how fast a wave moves.

The energy of a wave depends on its amplitude and frequency. Some waves lose energy as friction decreases the movement, causing the waves to die down after a while. Another important factor affecting energy is distance. A spherical wave is a wave emanating from a source in all directions, such as the sound waves created by clapping hands, or light coming from a lamp. Spherical waves generally obey the inverse square law—their strength decreases with the square of the distance from the source. This is because the energy per unit area decreases as a spherical wave moves outward—the spherical surface of the wave increases as it expands, but the energy stays the same, so there is less energy at each point. Consider sunlight, for example. An Earth-size planet orbiting two times farther from the Sun than Earth would receive four times less solar energy (since four is the square of two).

An interesting question can be raised about waves that travel in space, such as electromagnetic waves and the hypothetical gravitational waves. Space is virtually a vacuum, with sparsely distributed material such as hydrogen atoms and clouds of dust. If a wave moves through space, what exactly is "waving"? The question of the medium for electromagnetic waves arose in the 1860s, when the Scottish physicist James Clerk Maxwell (1831–79) predicted the existence of such waves and hypothesized that light is an electromagnetic wave. After Hertz produced and detected radio waves in 1888, confirming the existence of electromagnetic waves, scientists puzzled over what might be carrying these waves.

One proposal was that a pervasive material called the ether, or luminiferous (light-carrying) ether, was the medium. This hypothetical material filled the universe and, in order to serve its purpose, it had to have some peculiar properties. The stronger a medium's resistance to a wave's motion, the faster the wave can travel through the medium since the disturbance moves from point to point more quickly. Because light and other electromagnetic radiation is so fast, the ether had to be extremely rigid to this kind of disturbance. Yet the ether could not possibly affect matter, otherwise it would slow the motion of planets in their orbits, causing them to fall into the Sun.

Scientists searching for signs of a luminiferous ether could find no evidence that it existed. They finally gave up, especially after the German-American physicist Albert Einstein (1879–1955) introduced his special theory of relativity in 1905, which does not require any sort of ether. Physicists concluded that electromagnetic waves can propagate in empty space.

Electromagnetic waves consist of fluctuating electric and magnetic fields. Maxwell proposed that a disturbance in these fields can propagate, spreading out from the origin at the speed of light in a vacuum, denoted c. Certain motions of electric charges initiate these disturbances; for example, acceleration of a charge produces electromagnetic waves, as does the falling of an electron into a lower (less energetic) "orbit" in an atom. Although electromagnetic radiation has many properties of waves—for example, overlapping light beams can interfere, displaying light and dark bands corresponding to the addition or subtraction of the waves—Einstein proposed an alternative in 1905 in which light consists of particles called photons. (The year 1905, in which Einstein published a number of remarkable discoveries, is often referred to as Einstein's annus mirabilis, which is Latin for year of wonders.) Despite the difference between waves and particles, both descriptions of light are valid, a feature known as wave-particle duality.

Gravitational radiation or waves can also propagate through space. The waves are associated with disturbances in gravitational fields, similar to the electric and magnetic field fluctuations that comprise electromagnetic waves. Sometimes gravitational waves are referred to as gravity waves, but this can create confusion because physicists also use the term *gravity wave* to describe something else. In fluid physics, a gravity wave is a wave in which gravity acts to counter or restore the disturbance, as an ocean wave, in which gravity tugs at the rising crest and pulls it down again. The discussion in this chapter refers only to gravitational waves.

SPACE-TIME RIPPLES

Einstein published the general theory of relativity in 1916 after years of work. The theory uses complicated mathematical equations to describe the force of gravitation as a curvature of space-time, which consists of three dimensions of space and one dimension of time. This theory arose in part from efforts Einstein made to generalize his special theory of relativity, which applies only to conditions of constant velocity (see chapter 3), not acceleration such as that of an object falling in a gravitational field. In Einstein's view, the geometry of space-time was a more accurate and elegant depiction of gravitation than the concept of a force acting at a distance. The following sidebar discusses Einstein's view of the universe.

General relativity predicts that gravitational disturbances ripple through space-time at the speed of light. Using a flexible rubber sheet as a model of space-time, a massive object placed on the sheet will cause a depression, bending the fabric in the vicinity and causing smaller objects to roll toward it—this action represents gravitation. Now imagine what would happen if the heavy object bobbed up and down. This motion would create waves or ripples through the sheet. These waves are gravitational waves, created by the motion of the massive object.

Any object or system of objects having mass creates gravitational waves when it undergoes changes in velocity or in the arrangement of its components. These changes involve acceleration—a sudden movement or change in speed—of the object or its components. This is similar to a charged particle emitting electromagnetic waves when it experiences acceleration. But the emission of gravitational waves is complicated by the principle, known as Newton's third law, that every action has a reaction; each sudden movement of a mass is generally accompanied by an opposite movement, which tends to generate waves that cancel one another. Since the two movements are not in the same place, however—the two masses are separated by some distance—the waves are slightly different, so they do not cancel completely.

Einstein's Universe

The British physicist Sir Isaac Newton (1642–1727) conceived of gravitation as an attractive force exerted by masses on one another. According to his universal law of gravitation, the gravitational force is proportional to the product of the two masses and inversely proportional to the square of the distance between their centers. While Einstein acknowledged this formula's accuracy in most instances, the notion of a force reaching out over vast distances bothered him. Einstein's special theory of relativity embraced the speed of light as the maximum attainable velocity (see chapter 3), and he thought that objects should not be able to interact or project forces with any greater speed. The idea of a force instantaneously acting across the gulf of space failed to be appealing.

Instead, Einstein visualized gravity as geometry. While thinking about how this might work, Einstein pondered the path of a light beam in a gravitational field. As the biographer Walter Isaacson wrote in his 2007 book *Einstein,* "One solution might be to liken the path of the light beam through a changing gravitational field to that of a line drawn on a sphere or on a surface that is warped. In such cases, the shortest line between two points is curved, a geodesic like a great arc or a great circle route on our globe. Perhaps the bending of light meant that the fabric of space, through which the light beam traveled, was curved by gravity." Proceeding on this basis, and guided by other principles of physics, Einstein derived the complicated equations relating the curvature or warping due to the effects of gravity. Rather than considering gravitation as a force acting over a distance, Einstein viewed it as a bending of space-time, with the attraction of masses due to their following this curved path.

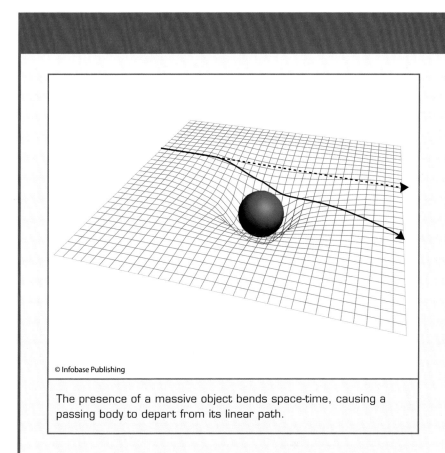

© Infobase Publishing

The presence of a massive object bends space-time, causing a passing body to depart from its linear path.

For an analogy, consider a rubber sheet with a heavy object placed in the center, which creates a depression, as shown in the figure. As a result of the curvature, other objects roll toward the heavy object—this is gravitation.

As Isaacson notes in his book, "Years later, when his younger son, Eduard, asked why he was so famous, Einstein replied by using a simple image to describe his great insight that gravity was the curving of the fabric of spacetime. 'When a blind beetle crawls over the surface of a curved branch, it doesn't notice that the track it has covered is indeed curved,' he said. 'I was lucky enough to notice what the beetle didn't notice.'"

Gravitational waves passing through an area would cause an alternating compression and stretching of space-time. This oscillation would vary the distance between a set of objects, for example. The amplitude of the oscillation depends on the wave. In general, the heavier the mass and the greater the acceleration of the source, the stronger the gravitational waves will be. The presence of matter does not block gravitational waves.

But gravitational waves are weak compared to electromagnetic waves. The force of electromagnetism is much stronger than gravitation. When matter is electrically charged, electromagnetism dominates its behavior instead of gravitation, and the canceling contribution of the reaction (due to Newton's third law) also decreases the amplitude of gravitational waves. Physicists quickly realized that detecting gravitational waves would be a challenge.

Knowing how gravitational waves theoretically behave would be a great help, but the equations of general relativity are extremely complex. Scientists at NASA, led by John Baker at the Goddard Space Flight Center in Maryland, recently used a supercomputer to simulate what would happen if two black holes collided and merged. Black holes, which are another prediction of general relativity, are extremely dense regions such as the remnants of collapsed stars where gravitation is so strong that nothing can escape (hence the term *black hole*). Supercomputers are fast computers that perform billions of calculations per second. The researchers' simulation produced complex waveforms that gravitational wave detectors can use to search for signs of black hole mergers in space. In the news release of April 18, 2006, announcing the result, Joan Centrella, the director of the Gravitational Astrophysics Laboratory at Goddard, said, "These mergers are by far the most powerful events occurring in the universe, with each one generating more energy than all of the stars in the universe combined. Now we have realistic simulations to guide gravitational wave detectors coming online."

HULSE-TAYLOR BINARY STAR SYSTEM: INDIRECTLY MEASURING A GRAVITATIONAL WAVE

Detecting the passing of gravitational waves from even violent events such as black hole collisions is difficult, as discussed in the following section. But researchers found indirect evidence of gravitational waves when they discovered a peculiar binary star system.

In 1974, Joseph H. Taylor, Jr., then a professor at the University of Massachusetts, and his student Russell A. Hulse were using the large radio telescope at Arecibo to search for pulsars. (For more information on the Arecibo Observatory, see the sidebar on page 6.) Pulsars, which had been discovered in 1967, are believed to be rotating neutron stars, the remnants of some types of supernova explosions. These dead stars do not shine, but they are small and compact, and emit pulses of radio waves as they rapidly spin. The researchers found a pulsar known as PSR1913+16 (*PSR* denotes *pulsar,* and the numbers signify the pulsar's location in the sky) located 21,000 light-years away. This pulsar was to prove significant for several reasons. It turned out to be the first binary pulsar ever discovered.

While studying the period of the pulsar's radio wave emissions, Hulse and Taylor noted a consistent variation, sometimes the pulses would arrive slightly sooner than usual, and sometimes slightly later. Careful measurements indicated that this effect persisted and was systematic, which indicated the perturbation of an object orbiting the pulsar. This effect is the same that was used later to discover the first extrasolar planets, as described in chapter 1. But in this case, the data indicated that the mass of the companion nearly equaled the pulsar. The companion, in other words, is a star. The Hulse-Taylor system is a binary star, orbiting around the system's center of mass with a period of about 7.8 hours. Each star has a mass about 1.4 times larger than the Sun's mass, and both appear to be neutron stars as indicated by the orbital mechanics. (The companion's pulses, if any, have not been detected, so it is not considered a pulsar, even though the system is called a binary pulsar.) The radius of these stars is about 12 miles (20 km). With so much mass in such a small space, pulsars have incredibly large gravitational fields—strong enough to squeeze all matter into neutrons, which is why they are known as neutron stars.

The high-gravity situation of a binary pulsar, along with the precision of the pulses, allows scientists to conduct important tests of gravitation. One of these tests involves gravitational waves. The small, dense stars should be generating gravitational waves as they swing around the system's center of mass. Although researchers cannot measure these waves directly, their emission should exert a measureable effect on the system. Gravitational waves have energy, as do electromagnetic waves and other kinds of wave. Because of the law of energy conservation—energy cannot be created or destroyed—the energy carried away by gravitational waves must come from the system. As a result, the gravitational potential energy of the system should be decreasing if it is emitting gravitational waves. If so, the orbiting stars should be getting closer, falling toward one another as they lose the energy carried away in gravitational waves.

Four years after the discovery of PSR1913+16, researchers had accumulated enough data to detect a small change in the orbital period. The rate of change agreed with what the theory predicted. Further observations have strongly reinforced this discovery. In a 2005 paper published in *Binary Radio Pulsars,* the astronomer Joel M. Weisberg and Taylor wrote, "The measured rate of change of orbital period agrees with that expected from the emission of gravitational radiation, according to general relativity, to within about 0.2 percent."

The steady orbital decay means that one day the stars will collide. Because gravitational waves are tiny and do not carry away much energy, the loss is small, and the system will not collapse for another 300 million years or so. In the meantime, PSR1913+16 provides strong but indirect evidence for gravitational waves.

ATTEMPTS AT DIRECT MEASUREMENT

Although the orbital decay of the Hulse-Taylor system gives scientists confidence that gravitational waves exist, it does not constitute proof. Suppose, for instance, that the system is losing energy by some other means. If so, then scientists have incorrectly attributed the loss to gravitational waves. This kind of mistake is unlikely because the rate of loss fits so well with the prediction of general relativity and gravitational wave physics, but such a coincidence is possible. Researchers would like to observe gravitational waves directly in order to remove all doubt of their existence.

Several other reasons motivate the need to observe gravitational waves. One benefit of measuring gravitational waves is that the study of these waves offers insight into the general theory of relativity. Scientists have a lot of faith in this theory because it has passed numerous experimental tests, yet there is a disturbing conundrum looming in the future. General relativity is one of the foundations of modern physics because of its accurate portrayal of gravitation and the universe, but general relativity does not explain everything. At the other end of the size scale—atoms and subatomic particles—is quantum mechanics, a set of principles that describe the behavior of these tiny objects. Quantum mechanics is another critical foundation in modern physics, and this theory has also passed numerous experimental tests, so scientists have as much faith in it as they do in general relativity. The problem is that these two theories do not mesh well. At the very core of these theories lie irreconcilable differences.

General relativity consists of principles and formulas by which scientists can predict the behavior of a set of objects interacting gravitationally. The formulas are deterministic: Given a certain set of circumstances, the objects will behave in a specific manner, as determined by the formulas. Quantum mechanics is not deterministic. Given a certain set of circumstances, the formulas of quantum mechanics only provide a set of probabilities that the par-

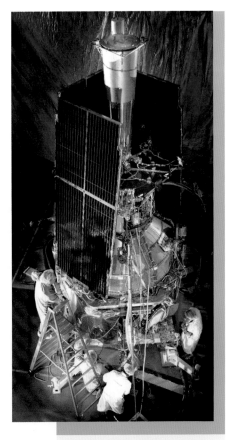

Gravity Probe B, launched in 2004, conducted tests of Albert Einstein's general theory of relativity. As a theory passes more tests, scientists become increasingly confident in its accuracy. *(Lockheed Martin Space Systems)*

ticles will behave in one way or another. For example, quantum mechanics might predict a 60 percent chance an atom will follow path *A* and a 40 percent chance it will follow path *B*. Six atoms out of 10 will follow *A*, but quantum mechanics cannot tell which ones out of the sample of 10 will do so. Heisenberg's uncertainty principle, which states that pairs of measurements such as position and momentum cannot be precisely determined simultaneously, also injects a certain amount of fuzziness into quantum mechanics.

Quantum mechanics and general relativity have no overlapping applications as yet since gravitation is weaker than other forces and is irrelevant on the tiny scale of quantum mechanics. But as researchers expand their horizons, particularly in high-energy experiments with particle accelerators, the time will come when these two theories meet, and at least one will prove to have limited applicability. If general relativity passes an increasing number of tests such as those conducted with gravitational waves, researchers will gain confidence that Einstein's theory is not the most likely to be limited.

Gravitational wave observations will also permit astronomical studies. This goal is discussed in the section "Gravitational Wave Astronomy" below.

Early attempts to measure gravitational waves did not succeed. In the 1960s, Joseph Weber at the University of Maryland suspended two aluminum cylinders, about 3.3 feet (1 m) in diameter and 6.6 feet (2 m) long, and monitored them for vibrations. Weber announced in 1969 that he had observed vibrations that signaled the passing of a gravitational wave, but other researchers pointed out that he had failed to account other sources of vibration, such as seismic waves.

What scientists needed was a more sensitive instrument that they could keep isolated from events such as earthquakes, or even the rumbling of a nearby garbage truck, which set off vibrations. A passing gravitational wave might come from black holes colliding at some vastly distant region of space, and the vibrations that these waves induce in the measuring instrument will be tiny. Other vibrations can easily drown out the desired signal, or it can mimic the passing of a gravitational wave and fool scientists into thinking they have observed one. The required instruments and facilities cost a great deal of money, but the National Science Foundation (NSF), a U.S. government agency that funds research, decided the effort was worth the expense. NSF is one of the most important funding agencies in the United States, and the following sidebar provides more information on this agency. Funds from NSF enabled researchers to build the Laser Interferometer Gravitational-Wave Observatory (LIGO).

Up to this point, LIGO is the largest project NSF has undertaken. The agency has invested about $300 million to build the facility and allocates $30 million a year to operate it.

National Science Foundation

World War II proved the value of science in the defense of the country and conducting military operations. Many important weapon systems, especially the atomic bomb that ended the war, were the products of scientific research funded by the government. After the war ended in 1945, officials realized that science could be put to a lot of important uses during peacetime as well. In 1950, Congress established the National Science Foundation (NSF) to promote and fund research that would lead to advances in citizens' health and welfare. The agency's headquarters is located in Arlington, Virginia.

According to the description of the LIGO project on NSF's Web site, "The National Science Foundation (NSF) provides funding for large, multi-user facilities that provide researchers and educators with access to the latest technological tools and capabilities. NSF also supports far-reaching areas of science and engineering that hold promise for breakthroughs that will enhance the nation's future in profound, and possibly unpredictable, ways. The Laser Interferometer Gravitational-Wave Observatory (LIGO) is an example of both."

Annual NSF budgets run about $6 billion. Much of this money gets distributed at college and university researchers, who compete for available funds. There is not enough money for all projects, so researchers who request funding must advance strong arguments in their favor. Projects such as LIGO that are doable and potentially rewarding will get funded.

LIGO: LASER INTERFEROMETER GRAVITATIONAL-WAVE OBSERVATORY

LIGO consists of two installations, one at Hanford, Washington, and the other near Livingston, Louisiana. Scientists from the California Institute of Technology and the Massachusetts Institute of Technology designed the facilities. Workers completed construction in 1999, and the gravitational wave detectors began operations in 2001.

The two installations have the same equipment but are separated by 2,000 miles (3,200 km). This separation helps researchers eliminate artifacts—signals that are not related to what the scientists are trying to measure. For example, suppose a slight Earth tremor shakes the facility in Washington, creating a "false positive" (such tremors occur frequently, although they are generally too small to be felt by humans). The facility in Louisiana will not be affected, or much less so. When researchers compare data from the two facilities, they can weed out local effects. A gravitational wave coming from space will have a broad front, easily extending across the distance separating the facilities and affecting both of them.

To detect the tiny oscillations expected from the passage of gravitational waves, researchers employ an instrument known as an interferometer. An interferometer makes precise measurements of distance or length by using the interference properties of light. Light is an electromagnetic wave, and when waves overlap, striking the same surface, they superimpose. Two waves of the same size and wavelength that are in phase—the crest of one wave occurs at about the same time as the other—add, making a combined wave twice as large as the two components. Waves that are completely out of phase—the crest of one wave occurs at the same time as the trough of another—cancel each other. When light waves interfere, the results are bands of bright light (where the waves add) and darkness (where the waves cancel).

Under ordinary conditions, light consists of waves in all phases and many different wavelengths, and any interference gets washed out. But when the waves are coherent and of a single wavelength, as in a laser beam, interference becomes noticeable. Beams of coherent light that hit the same place on a screen will interfere. For instance, suppose a laser beam is split, with half of the waves following one path and the other half following a different path. If the paths lead to the same destination, the interference will produce bands. Consider what would happen if one of the paths was slightly longer, for example, half a wavelength. The wave that took this path would arrive at the destination at a different time in its cycle, at the opposite phase (which is half a wavelength—see the figure on page 89). In this case, troughs and crests would superimpose, resulting in a dark band. Since the wavelength of light is so small—red light has a wavelength of 0.000028 inches (0.00007 cm)—interferometers can make precision measurements.

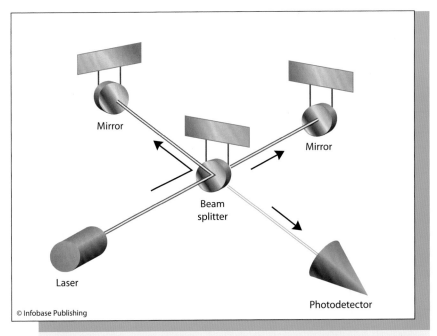

Mirror

Mirror

Beam
splitter

Laser

Photodetector

© Infobase Publishing

A beam-splitter sends part of the light down one arm of the instrument and the other part down the other arm, situated at a 90-degree angle. Mirrors at the end of the arms reflect the light, some of which meets at the photodetector.

LIGO's interferometers are L-shaped, having arms with a length of 2.4 miles (4 km). As illustrated in the figure, a beam-splitter (a partially silvered mirror) allows part of the light from a laser to travel one arm and part the other. The beams merge at a photodetector that measures light intensity. When a gravitational wave passes through the right-angled arms, the ripple will slightly decrease the length of one arm and increase the length of the other arm. The changes will affect the interference patterns measured by the photodetector. These changes are extremely small—about one hundred million times smaller than the diameter of an atom! Such precision requires nearly flawless equipment.

Other gravitational wave detectors have also been built. For instance, GEO 600, a German-British project, uses concepts similar to LIGO but with shorter arms—1,968 feet (600 m). Scientists often share data obtained with these instruments in the search for gravitational waves.

As of summer 2009, LIGO has not spotted any gravitational waves. Scientists are not discouraged because events emitting waves large enough to be detectable by these instruments are probably not very frequent. To upgrade LIGO's detection capability, NSF approved on March 27, 2008, a seven-year, $205 million project to enhance LIGO's sensitivity by a factor of 10. When completed, the installation of more sensitive instruments will permit LIGO to detect the gravitational waves from a much larger variety of events. In a press release issued by the California Institute of Technology on April 1, 2008, LIGO executive director Jay Marx said, "We anticipate that this new instrument will see gravitational wave sources possibly on a daily basis, with excellent signal strengths, allowing details of the waveforms to be observed and compared with theories of neutron stars, black holes, and other astrophysical objects moving near the speed of light."

GRAVITATIONAL WAVE ASTRONOMY

Routine detection of gravitational waves will play an important role in future astronomical studies. Gravitational wave detectors are already making a contribution, although at present only because of negative findings—they have shown that gravitational waves of a certain size do not accompany certain events.

On January 2, 2008, for example, the California Institute of Technology announced that LIGO had helped astronomers studying a particular *gamma-ray burst* (GRB). A GRB is a sudden, transient source of intense gamma rays, which are energetic electromagnetic waves having frequencies beyond even X-rays. GRBs last only a few seconds. On February 1, 2007, gamma-ray detectors in satellites picked up a burst of gamma rays coming from the direction of the massive galaxy M31, also called the Andromeda galaxy, which is about 2.5 million light-years distant. This GRB was named GRB070201 (from the date of detection— year 07, month 02, and day 01). To be detected at this great distance, the source of this radiation must have been a powerful emitter. In addition, because gamma rays have so much energy, their release signifies high-energy events. Many GRBs probably arise from violent collisions of massive objects.

Scientists working at LIGO checked to see if any gravitational waves accompanied GRB070201, as would be expected if the burst had originated from a collision of, say, two neutron stars or black holes. Since gravitational waves travel at the same speed as electromagnetic waves, the two types of waves should arrive at Earth at the same time if they come from the same source. But LIGO was silent. According to the news release of the California Institute of Technology, "Such a monumental cosmic event occurring in a nearby galaxy should have generated gravitational waves that would be easily measured by the ultrasensitive LIGO detectors. The absence of a gravitational-wave signal meant GRB070201 could not have originated in this way in Andromeda. Other causes for the event, such as a soft gamma-ray repeater or a binary merger [collision and merging of two objects] from a much further distance, are now the most likely contenders."

Soft gamma-ray repeaters may be neutron stars with decaying magnetic fields that release electromagnetic radiation in the form of gamma rays at the lower end of the gamma ray spectrum; these waves have less

Cross section of AURIGA, a gravitational wave detector in Italy that researchers expect to be useful in gravitational wave astronomy *(Tommaso Guicciardini/Photo Researchers, Inc.)*

energy than most gamma rays and are called soft. In the press release, the University of Florida physicist David Reitze said, "This is the first time that the field of gravitational-wave physics has made a significant contribution to the gamma-ray astronomical community, by searching for GRBs in a way that electromagnetic observations cannot."

If the upgraded LIGO equipment or other gravitational wave detectors are able to observe gravitational waves from many different sources, astronomers will be able to study a large number of events occurring in deep space. These studies are important not only to learn more about the nature of these events, but also in the study of the history and evolution of the universe. Distant objects and events provide essential clues on the evolution of the universe because looking into space is the same as looking back in time. Light takes time to cover the vast distances of the universe, so the light arriving at Earth represents objects and events that actually occurred a long time ago. For example, Andromeda galaxy is 2.5 million light-years away, which means light requires 2.5 million years to travel from there to here. When observers on Earth look at Andromeda, they see the galaxy as it was 2.5 million years ago, when it emitted the light that is just now reaching Earth.

Astronomers study the early universe by examining extremely distant objects. The big bang theory says that the universe was created about 14 billion years ago from an incomprehensibly dense point that exploded. Shortly afterward, astronomers theorize that the universe experienced a short period of extremely rapid growth, called inflation, and then continued to expand at a much slower rate. During inflation, a great deal of energy got converted into particles and radiation, and, according to the theory, the process generated an extraordinary quantity of gravitational waves. These gravitational waves would be detectable from all regions of the sky—a sort of distant background, billions of light-years away, which memorializes the inflationary period of the universe's history. The gravitational-wave background would be similar to the cosmic microwave background radiation, which was discovered in 1965 and is the remnant of the big bang explosion.

Inflation is an unproven theory, since evidence of the early universe is hard to come by; objects of the early universe are billions of light-years away and extremely faint. Detection of a gravitational-wave background would support the hypothesis, but only if astronomers were certain that the gravitational waves are due to inflation. In 2008, researchers at Case Western Reserve University in Ohio argued that there could be another source of this gravitational radiation. The university issued a news release on April 14, 2008, noting that "the key prediction of inflation theory is the presence of a particular spectrum of gravitational radiation. Detection of this spectrum was regarded among physicists as 'smoking gun' evidence that inflation did in fact occur, billions of years ago." But the Case Western Reserve University researcher Lawrence Krauss and his colleagues discovered another possible mechanism. In this complicated mechanism, the alignment of certain fields results in the emission of gravitational waves similar to those that would have been produced by inflation.

Future projects in gravitational-wave astronomy will include the search for this background. If it is found, researchers must decide which of the alternatives best account for the phenomenon. Perhaps another theory will emerge that offers the most accurate description of the evolution of the early universe. See chapters 5 and 6 for more discussion on this topic.

LISA: LASER INTERFEROMETER SPACE ANTENNA

LIGO's upgrade will help make gravitational waves easier to observe, but some researchers have proposed a new kind of detector. They want the new detector to observe gravitational waves from an ideal vantage point—in space.

A space-based gravitational wave detector avoids the tremors associated with ground movement. At low frequencies, especially frequencies below 1 hertz, these tremors overwhelm any possible gravitational wave no matter what precautions the ground-based detector takes. A space-based detector would also not be subject to the "noise" of modern civilization.

In 2001, NASA and the European Space Agency (ESA) agreed to work together on a project called the Laser Interferometer Space Antenna (LISA). This ambitious project aims to launch three identical spacecraft positioned 3,000,000 miles (5,000,000 km) apart in a triangular arrangement. Each vehicle will be at the vertex of an equilateral (equal-sided) triangle. The spacecraft will orbit the Sun at about one astronomical unit—the same distance at which Earth orbits the Sun—but will trail the planet at an angle of about 20 degrees. In other words, the angle formed by Earth, the Sun, and the center of LISA will be 20 degrees. This position is far enough from Earth to shield LISA from the planet's gravitational field, which would confound the observations, but close enough to permit easy communication from Earth to the vehicles.

Each of the three vehicles will carry identical instrumentation. The equipment includes lasers and sensors that form a gigantic interferometer, allowing a LISA vehicle to measure the distance between itself and the other two spacecraft with incredible precision. Current designs call for two "test masses" in each vehicle. The test masses are cubes 1.6 inches (4 cm) per side and are highly polished so that they reflect light. These masses are freely floating in the vehicle and act as mirrors for the interferometer.

Artist's rendition of LISA *(NASA)*

Suppose, for example, a gravitation wave passes through the space that LISA occupies. The faint rippling in space-time will slightly contract and expand the space between the vehicles as it rolls through the equilateral triangle. LISA's instruments will detect this rippling as changes in the interference bands of its interferometer. If LISA's equipment is sensitive enough, it will be able to detect gravitational waves from many different sources, including emissions from the earliest stages of the universe.

The design of LISA and the length of its equilateral triangle concentrate on detecting low-frequency gravitational waves. This design scheme complements rather than replaces LIGO and other ground-based detectors, which are endeavoring to find waves of higher frequency. Space- and ground-based detectors will study different facets of gravitational wave physics.

While LISA would avoid the tremors plaguing LIGO and similar detectors, the space-based gravitational wave detector must also overcome certain challenges. One of the biggest obstacles to success is vehicle drift. The solar wind, for example, might jostle the vehicles, and this motion would throw off the interferometer's sensitive measurements. Although the particles of the solar wind would not exert much force, it could be enough to spoil the precision required to detect the passage of tiny gravitational waves. Researchers will have to test LISA's equipment thoroughly.

If the technical challenges can be overcome, LISA would greatly boost the study and observation of gravitational waves. In the journalist Robert Irion's 2002 article in *Science,* the Montana State University astrophysicist Neil Cornish commented that the limitations of LIGO means that "you're really at the margins of detectability, struggling to try to dig signals out of the noise." He added, "People assume that's true for gravitational waves in general, but it is not at all the case for LISA. We have massive signal-to-noise, far better than for some optical instruments. We will see sources within the first hour of turning on."

NASA and ESA plan on launching the mission around 2019 or 2020. But LISA will cost millions of dollars, and budgets are tight and may get even tighter in the future. No one is certain if LISA will ever get off the ground—or even off the drawing board.

CONCLUSION

Space-based detectors such as LISA, if the project receives funding, and enhanced ground-based detectors such as LIGO will initiate the field of gravitational wave astronomy. Since gravitational waves travel through matter, all parts of the universe will be open to inspection; nothing will be obscured by clouds of dust and gas. Astronomers will be able to peer all the way to the very beginning of the universe.

The Hulse-Taylor binary system instills a firm belief that gravitational waves exist, and once scientists begin observing these waves, they can compare the observed properties with those predicted by general relativity. If the match is good, general relativity will have passed yet another test. Every time a theory clears a hurdle, scientists gain confidence in the theory. Powerful theories make many predictions, each of which offers the potential of failure. Philosophers such as Sir Karl Popper (1902–94) argue that any theory must make predictions and go "out on a limb" or it is not scientific. In science, theories must be falsifiable, or in other words, potentially proven wrong, if they are to have any use.

On these grounds, general relativity has become one of the most powerful theories in physics.

The study of gravitational waves will lead to tests of other theories besides general relativity. General relativity predicts gravitational waves, but if wave-particle duality applies to gravitation as well as electromagnetism—and there is no reason why it should not—then gravitational radiation should also exhibit particle properties. Particle physics is the domain of quantum mechanics. Gravitational radiation may offer a link between these two disparate ideas.

A particle of gravitation is known as a graviton. The main theory of particle physics, known as the Standard Model, describes the properties of particles and their interactions, or forces. According to the Standard Model, a special particle mediates or "carries" each of the forces—photons carry the electromagnetic force, particles called W and Z bosons carry the weak nuclear force, and gluons carry the strong nuclear force. Forces arise because objects exchange these particles. If this idea holds true for gravitation as well, then gravitation should have its carrier—the graviton.

But gravitation is so much weaker than the other forces that it plays little role in particle interactions. Physicists conduct experiments with particles by accelerating them to high speeds in gigantic machines called particle accelerators, crashing the particles together or into a target, then studying the debris. The violent collisions often create new and short-lived particles, allowing researchers to test their theoretical understanding of the process. With this technique, physicists have managed to describe the other interactions with a powerful theory involving quantum mechanics—the Standard Model—but gravitation has been left out. The notion of a graviton would fit within the framework of the theory and must exist if the theory is to be extended to all four forces. But no one has ever seen any sign of a graviton in these experiments.

Wave-particle duality indicates that physicists will be able to observe particles of gravitation once they begin detecting gravitational waves. But scientists are not sure if these gravitational particles will have the same properties of gravitons as suggested in quantum theories. If gravitational radiation travels at the speed of light, then gravitons should have no rest mass, similar to a photon. Discovering what other properties a graviton may have must await successful methods of observation.

Unification of quantum mechanics and gravitation is known as quantum gravity. Because of the differences in quantum mechanics and general relativity, theorists have failed to find any reasonable theory of quantum gravity. Yet the goal is extremely important. Scientists as well as detectives hate loose ends—unresolved mysteries—and the incoherence of the two great pillars of modern physics is a glaring loose end that researchers would love to tie up.

Gravitational waves are a frontier of science touching upon many different research topics. From the tiny world of particles to the width and breadth of the universe, gravitational waves will affect scientific research in astronomy and physics. When researchers finally manage to begin observing these waves, much of what is known may start coming together.

CHRONOLOGY

1687	The British physicist Sir Isaac Newton (1642–1727) formulates gravitational forces in his universal law of gravitation.
1864	The Scottish physicist James Clerk Maxwell (1831–79) predicts the existence of electromagnetic waves traveling at the speed of light.
1887	The American physicists Albert Michelson (1852–1931) and Edward Morley (1838–1923) perform an interferometer experiment that suggests ether, the hypothetical medium for electromagnetic waves, does not exist.
1888	The German physicist Heinrich Hertz (1857–94) produces and detects radio waves, electromagnetic waves with a longer wavelength than visible light.
1905	The German-American physicist Albert Einstein (1879–1955) proposes the special theory of relativity.

1916	Einstein proposes the general theory of relativity.
1918	Einstein discovers that the equations of general relativity predict the existence of gravitational waves.
1969	The pioneering physicist Joseph Weber prematurely announces the observation of a gravitational wave.
1974	Joseph H. Taylor, Jr., and Russell A. Hulse of the University of Massachusetts discover the first binary pulsar, PSR 1913 + 16, which will become instrumental in gravitational wave physics.
1978	Taylor and his colleagues note that orbital decay of PSR 1913 + 16 suggests the emission of gravitational waves, the first experimental evidence for these waves.
1990	National Science Foundation approves the Laser Interferometer Gravitational-Wave Observatory (LIGO) project.
1994	LIGO construction begins.
1999	LIGO facility inaugurated.
2001	LIGO begins operation.
	NASA and ESA agree to pursue the Laser Interferometer Space Antenna (LISA) project.
2008	The National Science Foundation approves an upgrade of LIGO to improve its sensitivity.

FURTHER RESOURCES

Print and Internet

California Institute of Technology. "Advanced LIGO Project Funded by National Science Foundation." News release, April 1, 2008. Available online. URL: http://mr.caltech.edu/media/Press_Releases/PR13123.

html. Accessed July 27, 2009. The California Institute of Technology announces that NSF is funding an upgrade of LIGO.

———. "LIGO Sheds Light on Cosmic Event." News release, January 2, 2008. Available online. URL: http://mr.caltech.edu/media/Press_Releases/PR13084.html. Accessed July 27, 2009. Researchers note the absence of gravitational waves associated with a gamma ray burst, which narrows down possible sources of the burst.

Case Western Reserve University. "Gravity Wave 'Smoking Gun' Fizzles, According to Case Western Reserve University Physics Researchers." News release, April 14, 2008. Available online. URL: http://blog.case.edu/case-news/2008/04/14/gravity. Accessed July 27, 2009. Researchers describe a mechanism besides inflation that could have produced gravitational waves early in the universe's development.

Einstein, Albert. *Relativity: The Special and the General Theory.* New York: Penguin, 2006. This translation and reprint of the 1916 original presents Einstein's simple and accessible explanation of his relativity theories.

Irion, Robert. "Gravitational Wave Hunters Take Aim at the Sky." *Science* 297 (August 16, 2002): 1,113–1,115. This article provides a discussion of the LISA project.

Isaacson, Walter. *Einstein.* New York: Simon & Schuster, 2007. This engaging biography of Albert Einstein includes materials such as personal letters that have only recently become available.

National Aeronautics and Space Administration. "NASA Achieves Breakthrough in Black Hole Simulation." News release, April 18, 2006. Available online. URL: http://www.nasa.gov/vision/universe/starsgalaxies/gwave.html. Accessed July 27, 2009. NASA scientists use a supercomputer to simulate the gravitational wave emitted during the collision of black holes.

National Science Foundation. "LIGO: The Search for Gravitational Waves." Available online. URL: http://www.nsf.gov/news/news_summ.jsp?cntn_id=103042. Accessed July 27, 2009. NSF offers a concise description of the LIGO project.

Parker, Barry. *Einstein's Brainchild: Relativity Made Relatively Easy!* Amherst, N.Y.: Prometheus Books, 2000. Physicist and prolific author Barry Parker offers an entertaining and accurate description of Einstein's special and general theory of relativity.

Thorne, Kip S. *Black Holes and Time Warps: Einstein's Outrageous Legacy.* New York: W. W. Norton, 1995. Thorne, a physicist who has made many contributions to the physics of gravitation and general relativity, explains Einstein's ideas on curved space-time and their fascinating consequences.

Weisberg, Joel M., and Joseph H. Taylor. "Relativistic Binary Pulsar B1913+16: Thirty Years of Observations and Analysis." In *Binary Radio Pulsars,* edited by F. A. Rasio and I. H. Stairs, ch. 3.

San Francisco: Astronomical Society of the Pacific, 2005. This book contains papers on binary pulsars written by experts.

Web Sites

Center for Gravitational Wave Physics. Available online. URL: http://cgwp.gravity.psu.edu/index.shtml. Accessed July 27, 2009. Based at Pennsylvania State University, the center is devoted to the study of gravitational waves and the pursuit of gravitational wave observations as an aid to astronomy. The Web site describes the center's projects and researchers.

Laser Interferometer Gravitational-Wave Observatory (LIGO). Available online. URL: http://www.ligo.caltech.edu/. Accessed July 27, 2009. This Web site offers news and information on the observatory's efforts to detect gravitational waves.

Laser Interferometer Space Antenna (LISA). Available online. URL: http://lisa.nasa.gov/. Accessed July 27, 2009. This Web site offers news and information on the plans to detect gravitational waves with a set of three spacecraft.

National Science Foundation (NSF). Available online. URL: http://www.nsf.gov/. Accessed July 27, 2009. The Internet home of NSF offers news and information on the research funded by this large government agency.

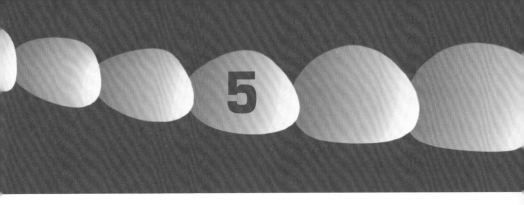

FORMATION AND EVOLUTION OF GALAXIES

The universe is filled with fascinating objects and phenomena. Astronomers have discovered incredibly dense stars called pulsars, which emit pulses of electromagnetic waves like a beacon, as well as planets beyond the solar system that orbit other stars (see chapter 1). The general theory of relativity predicts black holes—object so massive that not even light can escape them—for which scientists have found plenty of evidence, and gravitational waves, the ripples in space-time that researchers are still trying to observe directly (see chapter 4). All of these topics are worthy of study.

Another important topic in space and astronomy centers on what at first glance seems somewhat mundane. The universe contains billions of galaxies—vast congregations of stars, gas, and dust. A galaxy appears in telescopes as a misty cloud, and early astronomers, who were unsure of the nature of such an object, called it a *nebula* (Latin for cloud). Earth's solar system is contained in a large galaxy known as the Milky Way galaxy, which astronomers have studied from a much different vantage point than they have studied other galaxies—from within. Galaxies are not spread out evenly but instead tend to bunch together. No one knows the exact number of galaxies in the universe, but surveys and estimates suggest about 100 billion or more.

Distant galaxy *(NASA/ESA/Hubble)*

A galaxy-filled structure is not the only conceivable structure for the universe. For instance, stars could have been more evenly distributed rather than grouped into galaxies, and the universe was much smoother in its earliest stages. Scientists do not understand how galaxies formed, nor how they evolved into their present shapes. These mysteries underlie a great deal of research at the frontiers of space and astronomy, and the solutions to these mysteries will be critical factors in the scientific quest to understand how the universe came to be, and how stars, planets, and life took shape.

This chapter describes the basic issues of galaxy formation and evolution and discusses the current state of the science. Researchers have made impressive strides lately as they collect more data from orbiting telescopes and observatories. The new data give scientists the raw material with which to construct models to run on computers, simulating the slow but powerful processes that helped shape the universe into its present form.

INTRODUCTION

In 1920, an important scientific meeting took place in Washington, D.C., at the National Academy of Sciences. The primary participants, the American astronomers Harlow Shapley (1885–1972) and Heber Curtis (1872–1942), presented their differing opinions on the nature of nebulae (plural of nebula). The meeting is known as the Shapley-Curtis Debate, and their opposing viewpoints put a sharp focus on two contending schools of thought. Shapley argued that the nebulae astronomers observed in their telescopes were clouds of dust and gas within the galaxy (the Milky Way galaxy, which consists of a disk about 100,000 light-years in diameter), which in his view was the extent of the universe. Curtis believed the nebulae represent other, faraway galaxies.

Neither scientist made a convincing case during the debate. Shapley claimed the nebulae were much closer than Curtis thought, but at the time, astronomers could not be certain of the distances involved. As is generally the case in science, the debate was not resolved with words but with experiments and observations. In 1924, the American astronomer Edwin Hubble (1889–1953) established that most of the nebulae lie at vast distances from Earth. Curtis had been right—nebulae are distant galaxies.

Determining the distance of remote objects is not at all simple. Nearby stars show parallax—a change in apparent position as the observer shifts viewpoints—which astronomers can use to calculate the distance to the closest stars. But the parallax of all but the nearest stars is too small to measure.

In 1912, the American astronomer Henrietta Leavitt (1868–1921) discovered an important relationship in a class of stars known as *Cepheid variables* (named after Delta Cephei, the first such star she found). In a variable star, *luminosity*—the rate of emitted radiation—changes, so the star varies in brightness. A periodic variable star exhibits a regular cycle of changes, cycling from low to high in a consistent period of time. Leavitt discovered that Cepheid variables exhibit a relationship between the period and luminosity—the longer the period, the greater the luminosity.

The periods of Cepheid variables range from about one to 100 days. Some of the stars are bright enough to be seen in other galaxies, out to about 100 million light-years. Astronomers take advantage of the period-luminosity relationship by measuring the period of a Cepheid variable, and then calculating the luminosity. Luminosity reveals how much radiation a star is emitting, so astronomers can determine how bright the star will be at various distances. The radiation that an observer measures per unit area decreases with the square of the distance from the source, so a star with a certain luminosity will look four times dimmer when it is twice the distance of another star with the same luminosity. Knowing the luminosity, astronomers can measure apparent brightness, which gives the distance.

Hubble's 1924 discovery relied on Cepheid variables. After using a then new 100-inch (2.5-m) telescope at Mount Wilson in California to study faint Cepheid variables in the nebula that became known as Andromeda galaxy, Hubble calculated it was nearly a million light-years away. This is a staggering distance, well beyond the confines of the Milky Way galaxy, given the estimates of its size by Shapley and other astronomers. Modern measurements of the Milky Way put the diameter at about 100,000 light-years, and the distance to Andromeda galaxy is about 2,500,000 light-years.

Because individual stars are difficult to detect and measure at great distances, the use of Cepheid variables in distance measurements is limited. But scientists soon found an alternative method. Hubble announced in 1929 that the universe was expanding. He based this claim on observations that light from distant galaxies is shifted toward the red end of the spectrum—the *redshift*. Because of the Doppler effect, when a source of light and an observer are in relative motion, the frequency of the light is shifted (see the sidebar on page 10). Light from approaching objects is higher in frequency (shifted toward blue, the upper range), and light from receding objects is lower (shifted toward red, the lower range). Astronomers use specific spectral lines, representing elements and compounds, to gauge frequency shifts in the spectra of astronomical bodies.

Hubble formulated a law, known as *Hubble's law,* describing a proportional relationship between the velocity, v, with which an object is receding, and its distance, d. The relationship is mathematically expressed as $v = Hd$, where H is the constant of proportionality. Objects at greater distances have larger redshifts (which means they are receding at faster velocities). Determining the value of *H*, often called the *Hubble constant,* gives the scale of the known universe.

To find the value of *H*, Hubble and other astronomers have measured redshifts of galaxies and computed their distances with methods such as Cepheid variables, or certain supernova properties (see chapter 6). Such data allow researchers to find the best fit for *H*. One of the main goals of the *Hubble Space Telescope,* named in the astronomer's honor, was to find the value of this constant. The numerical value depends on which units of measurement are used, but the present value of the constant suggests that the most distant objects that have been observed are about 13 billion light-years away.

Astronomers have found and studied numerous galaxies and other objects with increasingly large telescopes such as the 33-foot (10-m) reflecting telescopes of the W. M. Keck Observatory atop Mauna Kea in Hawaii, and the 7.9-foot (2.4-m) mirror in the orbiting *Hubble Space Telescope.* Keeping track of all these objects has been a burden, even in early times, when telescopes were not so powerful. The French astronomer Charles Messier (1730–1817) compiled a list of more than a 100 prominent objects, which are given a number corresponding to the place on the list, preceded by the letter *M*. For instance, the Andromeda galaxy is M31. The New General Catalog, first compiled and published by the astronomer John L. E. Dreyer (1852–1926) in 1888, lists thousands of objects, designated by the letters NGC and then a number. Andromeda galaxy is NGC 224. Astronomers refer to other, less prominent galaxies by position or date discovered.

Many of these galaxies have specific shapes. Hubble was the first astronomer to begin classifying galaxies based on shape when he noticed many galaxies were disks with spiral arms, such as the Andromeda galaxy, and others had a globular or elliptical shape. Astronomers continue to use this broad classification scheme, with three main categories: spiral, elliptical, and irregular (which lack structure or regular features). With so many billions of galaxies in the universe, no one can ascertain for certain how many belong to each category, and estimates based on surveys vary due to the limited sample size. According to the ninth edition of Michael Zeilik's textbook *Astronomy—The Evolving Universe,* "A complete survey of a region out to 30 Mly [million light-years] showed that 34 percent of the galaxies in this volume are spirals, 12 percent are ellipticals and 54 percent are irregulars."

M81, a spiral galaxy *(NASA/JPL-Caltech/ESA/Harvard-Smithsonian CfA)*

Properties other than shape are also important in the study of galaxies, as discussed in the following two sections. But the very idea that galaxies can be sorted into classes suggests that they may have experienced similar formation and evolutionary processes. The classification scheme gave researchers at the frontier of space and astronomy the impetus to explore the issue further.

BIRTH OF THE UNIVERSE AND THE FORMATION OF GALAXIES

According to the big bang theory, the universe began in a violent explosion of a small and incredibly dense point about 14 billion years ago. Evidence that supports this theory includes the expansion of the universe, which is a continuation of the initial explosion. Another important piece of evidence is the cosmic microwave background radiation. This radiation fills the sky in all directions with a steady, low background of radiation. When astronomers examine the sky with a telescope that detects this radiation, they see a uniform glimmer everywhere they look. Such a background is highly unlikely to be due to a local source; the best explanation is that it is the remnant of the violent explosion that created the universe.

The uniformity of the background radiation implies that the young universe was smooth and its contents evenly distributed. Considering the violence of its birth, the smoothness of the young universe is difficult to understand. Scientists have theorized that a period of rapid expansion occurred early in the universe's history. This period, called inflation, which the theoretical physicist Alan Guth at the Massachusetts Institute of Technology proposed in 1980, solves several troubling aspects of the big bang theory. Inflation would be responsible for the relative flatness of space (apart from gravitational fields, as visualized in general relativity) as well as the uniformity of the early universe. In this scenario, the universe was tiny following its birth, at which time it could develop a sort of equilibrium, and the period of extremely rapid expansion that ensued was so fast that it maintained its uniformity. In other words, expansion happened too quickly for the components to get badly disarranged.

Cosmologists have searched for evidence of inflation but have seen no firm signs of it. This hypothetical period early in the universe's history should have produced plenty of gravitational waves—the subject of chapter 4—and researchers hope that direct observation of these waves may aid in the study of the early universe.

But the background's *isotropy*—sameness in all directions—poses a question. Scientists wonder how the universe evolved from the uniformity of its earliest stages to the lumpy distribution of matter seen today. Matter somehow collected into groups, forming stars and galaxies.

To study this problem, scientists have employed two satellites, each containing extremely sensitive instruments, to map the background radiation with precision. NASA launched the *Cosmic Background Explorer (COBE)* satellite on November 18, 1989. This satellite contained several instruments, including a differential microwave radiometer to look for minuscule variations in the intensity of the background radiation. *COBE* found *anisotropy*—a lack of uniformity—although the differences in different directions were slight. To further explore these differential measurements, the *Wilkinson Microwave Anisotropy Probe (WMAP)*, named in honor of the cosmologist David T. Wilkinson

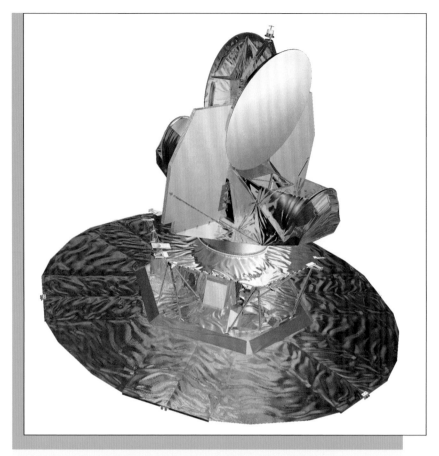

Artist's drawing of the *Wilkinson Microwave Anisotropy Probe (WMAP)*, a satellite that has provided a lot of valuable astronomical data *(WMAP)*

(1935–2002), rocketed into orbit on June 30, 2001. *WMAP* has found tiny patterns in the radiation that scientists do not fully understand, but these fluctuations confirm the lack of absolute uniformity.

Out of this slightly clumpy distribution came the galaxies. Although researchers do not understand the details, a simple idea to explain galaxy formation is that gravitational attraction acted to pull the clumps together. Starting with the small variations in the early universe, perhaps caused by quantum fluctuations, areas with slightly more mass began attracting surrounding particles, forming distinct centers.

Researchers who are working on this problem train their instruments on extremely distant objects. Because of light's finite speed, it takes a long time to travel intergalactic distances. Astronomers incorporate light's speed into their unit of measurement—the light-year. Light from an object 10 million light-years away, for example, shows the state of that object 10 million years ago, when light first began its journey. By looking far into space, astronomers look back in time.

In 2003, researchers using the *Hubble Space Telescope* identified numerous objects that are so distant they may be among the earliest galaxies to form. Light from these distant objects, about 13 billion light-years away, show the universe in an early period of its development. This period, about a billion years following the big bang, corresponds with a phase at which stars and galaxies began to form. The activity followed an era in the universe's history that astronomers refer to as the Dark Ages, during which the universe cooled off enough for protons and electrons to form atoms of hydrogen, which created clouds that acted somewhat like a fog, impeding the transmission of light. Current opinion has the onset of the Dark Ages around 300,000–400,000 years after the big bang, and ending about half a billion years later. In a press release issued by the Space Telescope Science Institute on January 9, 2003, the Arizona State University astronomer Rogier Windhorst noted, "The objects we found are in the epoch when the universe started to produce stars in significant numbers—the hard-to-find young galaxies. These galaxies are at the boundary of the directly observable universe."

These objects are difficult to study because they are at the limits of present technology's capacity to resolve and identify them. But the researchers Rychard Bouwens and Garth Illingworth at the University of California, Santa Cruz, recently used the *Hubble Space Telescope* to take a survey of these distant objects. They found hundreds of galaxies shining brightly at about the end of the Dark Ages, but only a few farther back. In a press release posted at ScienceDaily on September 13, 2006, Illingworth said, "The bigger, more luminous galaxies just were not in place at 700 million years after the big bang. Yet 200 million years later there were many more of them, so there must have been a lot of merging of smaller galaxies during that time." This research narrows down the time frame in which researchers should seek "baby" galaxies with improved instrumentation in the future.

GALAXY EVOLUTION

Researchers are interested in how galaxies developed their specific shapes once they began to form. As Hubble noted, galaxies with definite shapes tend to be either disks with spirals or elliptical without spiral arms. The question is how they got this way.

Astronomers are most familiar with the galaxy in which the solar system resides—the Milky Way galaxy, named from the band of light, called the Milky Way, which stretches across the night sky. This light, best observed far from the pollution and competing lights of a large city, represents numerous stars lying in the disk of the galaxy. The composition and dynamics of the galaxy suggest that it is about 13 billion years old.

Discovering the shape of the galaxy took a little detective work. From their inside perspective, astronomers have made observations of many different wavelengths of electromagnetic radiation, especially infrared, which penetrates dust clouds that otherwise obscure parts of the galaxy. Held together by gravity, stars journey around the center of the galaxy in orbits similar to the dynamics of the planets in the solar system. Researchers measure Doppler shifts to discover how fast stars are moving, and along with their position, this information helps scientists to infer the shape of the galaxy and its dynamics. Astronomers have concluded that the Milky Way galaxy is a large spiral galaxy with four spiral arms. The Sun is located about 25,000 light-years from the center of the galaxy, in or near one of the spiral arms.

How do the spiral arms form? The existence of many spiral galaxies indicates that the process or processes are not rare. But the dynamics of thin, spinning objects can be complex and unstable. Researchers are not certain what keeps spiral galaxies from quickly growing disorganized.

One approach to understanding spiral galactic structures is the density wave model. A density wave is a wave created by alternating compressions and rarefactions passing through a material. Sound waves are examples of density waves. The density wave model of spiral galaxies attributes the arms to density waves propagating around the galactic disk. Such waves could be stable and persist, causing the compression of the interstellar material. As an explanation, the density wave model works fairly well, but a glaring weakness in the theory is that no one knows what could get such waves started in the first place. This issue will be the subject of much more thought and work in the coming years.

The evolution of a galaxy takes far too long to watch, so astronomers can never expect to observe one progress from stage to stage. Instead, researchers look for young galaxies in the process of development, attempting to find examples at each stage. Assuming the process is similar in all or most galaxies, astronomers may be able to use these "snapshots" to piece together the mechanisms underlying galaxy evolution.

Early astronomers wondered if spiral galaxies may be a more advanced stage of an evolution that began with elliptical shapes. In this idea, elliptical galaxies were formed by gravitational attraction and then slowly evolved into spiral ones. But the evidence argues against this idea. One problem is that the angular momentum of a system is conserved. A system such as a galaxy, if left on its own, would retain its spin and its shape, so it is difficult to understand how one shape could evolve into another.

Another problem involves the color of a galaxy—the spectrum of its electromagnetic emissions—that involves the age of the stars that the galaxy contains. Stars progress from birth, which astronomers believe occurs out of swirling clouds of interstellar gas, to a shining maturity with the energy supplied from nuclear reactions in their core. Then, once the nuclear fuel is spent, death, which may be a spectacular explosion for stars of large mass, releases gas into interstellar space for the next generation. The lifetime of a star depends on its mass, although the relation may seem counterintuitive—small stars tend to last longer, despite less mass, because they tend to burn fuel more slowly. This slow rate results in lower temperatures, which give them a more reddish color than large stars that live in the "fast lane" and burn fuel quickly. The high temperatures of these high-mass, short-lived stars give them a bluish color, which is more energetic than red because it has a higher frequency.

Spiral galaxies tend to be bluer than elliptical galaxies. This indicates that spiral galaxies contain a greater population of young stars, for these short-lived stars must have been born in the recent past. (Small stars such as the Sun live for billions of years, while high-mass stars live for 100 million years or so, and much less for extremely large stars.) Many of the younger stars in spiral galaxies seem to be located in the spiral arms.

If elliptical galaxies belonged to the first stage of galactic evolution, then they should not be older but younger since they would represent

GALEX: Galaxy Evolution Explorer

Launched on April 28, 2003, *GALEX* is a NASA satellite designed to survey the universe from its orbit, 430 miles (690 km) above Earth. *GALEX* makes its observations in the ultraviolet portion of the electromagnetic spectrum—radiation that is slightly higher in frequency than visible light. The satellite's mission was originally planned to last 29 months, but in 2006 NASA scientists recommended an extension. In 2008, *GALEX* reached its fifth anniversary and is still going strong. The Jet Propulsion Laboratory (JPL) in Pasadena, California, operates and manages the mission.

As indicated by its name, the goal of this satellite centers on the identification and study of galaxies. From 2003 to 2008, *GALEX* imaged 500,000,000 objects. Many of these objects are faraway galaxies. In spiral galaxies, the arms tend to have a lot of gas, dust, and young stars, and they show up only faintly with optical instruments tuned to visible light. With its sensitive instruments concentrating on ultraviolet frequencies, *GALEX* can spot spiral arms in distant galaxies and their bluish inhabitants much more readily.

GALEX has accumulated so many observations and so much data that researchers are just getting started analyzing it. In a press release issued on April 28, 2008, by JPL and the California Institute of Technology, Chris Martin, the principal investigator for the Galaxy Evolution Explorer Mission, said, "Frankly we have only begun to scratch the surface of this vast data set. Astronomers will be mining the telescope's data archive for the next decade."

young galaxies in this scenario. But perhaps the evolutionary relationship goes the other way. Perhaps spirals evolve into ellipticals, just the opposite of what early astronomers guessed. In addition to the star populations, evidence for this idea comes from new observations from a satellite called *GALEX—Galaxy Evolution Explorer.* The sidebar provides more information on this important orbiting observatory.

GALEX surveys have turned up millions of galaxies. Scientists who are studying this enormous data set have recently found evidence of a concept called the nurture theory, in which young spiral galaxies evolve into older elliptical galaxies. In a Jet Propulsion Laboratory press release issued on November 14, 2007, *GALEX* researchers announced the discovery of "teen-aged" galaxies, making the transition from young to old. The *GALEX* investigator Chris Martin said, "Our data confirm that all galaxies begin life forming stars. Then through a combination of mergers, fuel exhaustion and perhaps suppression by black holes, the galaxies eventually stop producing stars."

COLLISIONS ON A GRAND SCALE

The mergers Martin referred to involve impressively large collisions. When galaxies collide, the results can be tumultuous. And due to gravitational attraction, galaxies can move around quite a bit.

Gravitational forces bound stars in their galaxy. Galaxies contain huge amounts of matter, and stars in the Milky Way galaxy sweep across vast orbits around the galaxy's massive center. The study of the dynamics of stars and star clusters has given researchers reason to believe that galaxies contain more matter than is visible, leading to the concept of dark matter, one of the subjects of chapter 6. All of this mass, both visible and dark, exerts a gravitational effect on other nearby galaxies as well as stars within the galaxy itself. Galaxies tend to bunch together in clusters. Clusters are prominent in wide-field surveys, such as images astronomers have obtained with the *Hubble Space Telescope.*

The Local Group, which is the name astronomers have given to the cluster containing the Milky Way galaxy, contains about 30 galaxies. Dominating this cluster are two large spiral galaxies—the Milky Way galaxy and M31, the Andromeda galaxy. Most of the other galaxies of the local group are small ellipticals.

Receding galaxies have redshifted spectra, and Hubble's law formulates the general expansion of the universe. But M31 does not have a redshift but rather a *blueshift,* which means the galaxy is moving toward the Milky Way galaxy rather than away from it. The Andromeda galaxy is one of the few galaxies that is getting closer, although at the pace it is

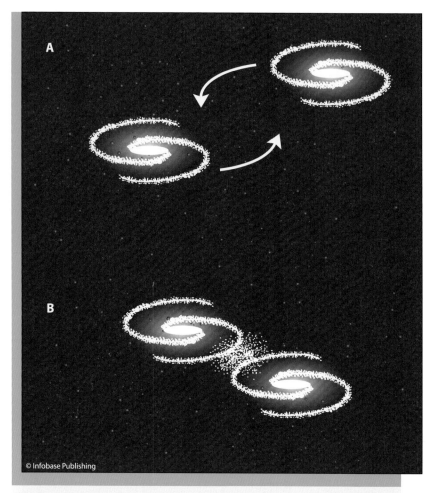

© Infobase Publishing

(A) Two spiral galaxies approach each other; (B) the collision disrupts the spiral arms.

traveling—about 90–180 miles/second (150–300 km/s), it will not cover the 2.5 million light-years separating it from the Milky Way galaxy any time soon. Yet despite the expansion of the universe, gravitation is pulling M31 and the Milky Way galaxy together.

The widest surveys of the galaxy show that even clusters of galaxies are clustered, creating what astronomers call superclusters. With all this motion, galaxies can travel complicated routes—giving them plenty of opportunities to run into one another.

Colliding galaxies are not like two solid objects such as cars slamming into each other. Stars in a galaxy are usually separated from their neighbors by vast distances, usually about 1 million times their diameter, although galactic centers tend to be much denser. (In contrast, galaxies are separated on average by much smaller factors of their diameters—the gulf between galaxies averages only about 100 times the typical galactic diameter.)

But gravitational interactions between colliding or closely approaching galaxies can strongly rearrange their shape. As illustrated in the figure on the left, interactions between two spiral galaxies can distort the arms, as well as initiating activity that leads to new star formation from "tidal" forces—tugs of gravitation that set clouds of gas and dust into motion, eventually swirling and condensing into a new star (and possibly a planetary system).

Astronomers believe that large elliptical galaxies probably arose from the merger of smaller galaxies. As a large galaxy grows, it attracts smaller galaxies nearby, absorbing this material and growing even larger. Scientists refer to this process as galactic cannibalism. Giant elliptical galaxies generally have several properties suggesting "merger and acquisition" activity. The position of these galaxies, which tends to be near the center of a cluster, and their motion, along with the presence

Shape-distorting gravitational interaction between two galaxies *(NASA, ESA, and Hubble Heritage Team [STScI/AURA])*

Artist's conception of a large black hole swallowing part of a star
(NASA/JPL-Caltech/Tim Pyle [SSC])

of clumps of stars scattered around their core, suggests giant ellipticals have accumulated lesser galaxies.

Using the *Hubble Space Telescope,* astronomers recently captured an image of NGC 1132, a giant elliptical galaxy at a distance of about 320 million light-years in the constellation of Eridanus. A press release posted on February 7, 2008, at ScienceDaily stated that researchers believe NGC 1132 "is, most likely, a cosmic fossil—the aftermath of an enormous multi-galactic pile-up, where the carnage of collision after collision has built up a brilliant but fuzzy giant elliptical galaxy far outshining typical galaxies." In the image, the press release noted that "NGC 1132 is seen surrounded by thousands of ancient globular clusters, swarming around the galaxy like bees around a hive. These globular clusters are likely to be the survivors of the disruption of their cannibalized parent galaxies that have been eaten by NGC 1132 and may reveal its merger history."

BLACK HOLES AND GALAXIES

Galaxies are not the only objects in which mergers and collisions spur growth. At the heart of many galaxies lies what may be an extremely large black hole.

A black hole is a strange object. The death throes of large stars often involve an explosion called a supernova, but much material is left behind, slammed inward by the departing gases. The remnants collapse under gravitational forces since there are no longer any nuclear reactions in the core to produce heat and radiation that offset these forces. If the mass is great, the material gets squeezed into such a small volume that a neutron star forms (the pressure squeezes protons and electrons into neutrons, so the whole star is a sphere of neutrons). An even greater mass results in so great a gravitational force that nothing can halt the collapse; the volume approaches zero, and the density approaches infinity. This process generates a black hole. Under these bizarre conditions, nothing can escape the

Artist's conception of a quasar *(NASA/JPL-Caltech)*

Quasars—the Heart of a Galaxy

In 1960, Allan Sandage, an astronomer at the Mount Wilson and Palomar Observatories in California, and his colleague Thomas Matthews identified a faint object corresponding to a previously discovered radio source in the constellation Virgo. Thousands of other such objects, known as quasi-stellar radio sources or quasars, have since been found. Despite the name, many of these objects are not powerful radio emitters, but emissions include a great deal of energy in the ultraviolet and X-ray portions of the spectrum.

When they were initially discovered, quasars puzzled astronomers. Many quasars vary in brightness rapidly, sometimes changing every few days. This means quasars must be small; if the objects had extensive volumes, a change propagating through them would require a lot of time (recall that the speed of light, though fast, is finite). But astronomers discovered quasars have extreme redshifts, indicating a distance in the billions of light-years. In order to be detectable at such a range, quasars must be some of the most luminous objects ever discovered, with the output of an average galaxy—but packed in a region that is thousands or millions of times smaller.

Current models of quasars suggest they are huge black holes at the centers of active, young galaxies, as illustrated enormous gravitational field. Even light is trapped, so the object will appear completely black.

Since no one has ever examined a black hole—they cannot be observed directly because they capture any light that strikes them—the dimensions are theoretical, predicted by the complicated equations of the general theory of relativity. The German physicist Karl Schwarzschild (1873–1916) calculated the minimum volume that a given mass must have in order to have a gravitational field strong enough to trap light. For an object with the same mass as the Sun, a black hole would be a

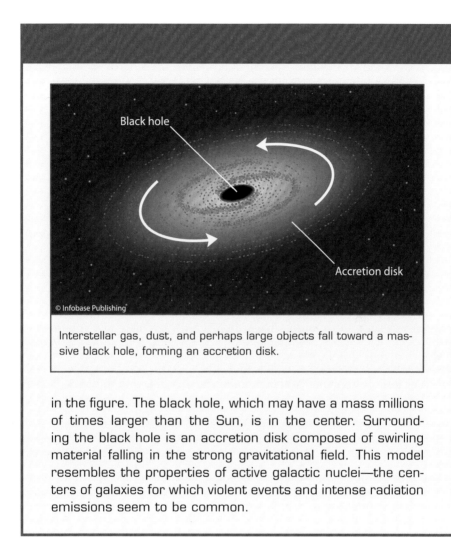

Interstellar gas, dust, and perhaps large objects fall toward a massive black hole, forming an accretion disk.

in the figure. The black hole, which may have a mass millions of times larger than the Sun, is in the center. Surrounding the black hole is an accretion disk composed of swirling material falling in the strong gravitational field. This model resembles the properties of active galactic nuclei—the centers of galaxies for which violent events and intense radiation emissions seem to be common.

sphere with a radius of about 2 miles (3.2 km). Compare this size to the Sun's radius of 430,000 miles (700,000 km).

Although black holes do not reflect any light, they exert effects on other objects. No one is certain exactly what happens to the mass in a black hole, but its gravitational field remains. Matter such as interstellar gas or dust—or a planet or star if it gets too close—gets drawn toward the black hole. The intense gravitation accelerates particles, stretching or pulling them and heating them up, stripping electrons and creating ions. Accelerating ions emit detectable amounts of electromagnetic radiation.

Much of this radiation is energetic enough to be in the X-ray range. As-tronomers have found intense X-ray sources that are small and do not show up in visible light—excellent black hole candidates. As matter falls into the black hole, its mass increases.

Other strong black hole candidates are *quasars.* The term *quasar* derives from the term *quasi-stellar radio source,* since radio astrono-mers first identified these objects with their radio telescopes, and re-searchers subsequently found faint, starlike points of light associated with them. Thousands of quasars have been found. Redshifts indicate that these objects are vastly distant, up to billions of light-years away. These small, remote objects must emit vast amounts of energy to be detectable, much of which is in the high-frequency range, such as ul-traviolet and X-rays, rather than radio waves. Astronomers believe that they are probably black holes at the centers of active, young galaxies. The sidebar on page 130 provides more details on quasars.

The idea of a black hole in the center of a galaxy is not confined to quasars. Astronomers have discovered extremely fast rotations of matter near the core of the Milky Way galaxy and in other galaxies, as revealed by Doppler shifts of the objects. These high velocities strongly suggest the existence of an extremely massive object in the galaxy's core. A dense clustering of stars might explain the mass, and although as-tronomers have failed to observe them, their light may be obscured by thick clouds of dust and gas. But the density of the galaxy's core suggests the presence of a huge black hole with a mass about 4 million times that of the Sun. As gas, dust, planets, and even stars fall into the black hole, its mass grows even larger.

If galaxies contain massive black holes in their centers, astronomers wonder how these black holes got there. Perhaps the black holes ap-peared first as the galaxy began to form and condense in its early stages of birth; alternatively, perhaps the galaxy came first, then the black hole arose as more and more matter collected at the core and collapsed.

A recent study using the Very Large Array, a group of 27 radio tele-scopes positioned in a plain near Socorro, New Mexico, found evidence that black holes may have come first. Earlier research indicated that the ratio of two specific masses in a galaxy—that of the black hole in the core and the mass of the stars in its central bulge—formed a ratio that is almost identical for many of the galaxies that have been measured. (The bulge's mass is about 1,000 times greater.) Chris Carilli, an astronomer at the National Radio Astronomy Observatory, and a team of researchers studied young galaxies, which are billions of light-years distant. They found that the black holes in these young galaxies are comparatively greater than in older galaxies; in other words, the bulge-to-black hole mass ratio is less. This would mean that the black holes probably appeared first, and the ratio decreased over time as galactic matter built up around the black hole. ScienceDaily posted a news release on January 7, 2009, announcing the results. In the release, Carilli said, "It looks like the black holes came first. The evidence is piling up."

SIMULATING GALACTIC EVOLUTION WITH COMPUTERS

Processes and mechanisms that act on galactic scales, stretching across thousands of light-years, are far too gradual to be observed over the course of a human lifetime (or many lifetimes). Astronomers search for galaxies in various stages of evolution, providing a peek at a certain phase in the cycle, but it is often difficult to deduce the order in which these stages occur. In order to investigate galactic evolution, scientists turn to computers to speed up the process.

Computer programs incorporate observations and principles or theories in physics to simulate processes that scientists have yet to observe, or lack the means or instruments to do so. Scientist distill what they believe are the essential elements and interactions of the process, building what is known as a model, which simplifies the situation as much as possible. (But if the model is too simple, then it is worthless, or even worse, misleading.) Astronomical observations and measurements yield important data that help researchers decide the values of the variables, or parameters, of the model. The model includes mechanisms by which the objects being simulated behave and evolve, which in the case of galaxies and cosmology is principally the general theory of relativity, formulated by the German-American physicist Albert Einstein (1879–1955) in 1916. Scientists use models and simulations to help them understand complex processes and to guide and predict future experimentation.

A model is only as good as the data and theory it uses. Inaccurate measurements can ruin the model, causing it to behave in ways that do not mimic reality. The theory must also be correct and applicable if the simulation is to yield useful results. In the case of cosmological simulations, researchers are fortunate in that scientists have much confidence in the general theory of relativity (see chapter 4).

Recent simulations have focused on black holes. In one of the simulations, the Carnegie Mellon University researcher Tiziana Di Matteo and her colleagues built a model to investigate the evolution of galaxies with black holes in the core. As galaxies began to form in the crowded conditions of the early universe, they would have often collided and merged. Di Matteo and her research team simulated this process and found that the collision resulted in the black holes coalescing. The massive combination attracted nearby gas and dust, becoming a quasar. Energy emissions of the powerful quasar heated and drove off other gases in the vicinity, which cut off the supply of incoming matter. After a while, the black hole no longer consumes matter, so it becomes quiet.

The "wind" that drives off the gases and shuts off the black hole's input is the key to this process, according to the model. In a Carnegie Mellon University news release dated February 9, 2005, Di Matteo's colleague Volker Springel at the Max Planck Institute for Astrophysics in Germany said, "This process inhibits further black hole growth and shuts off the quasar, just as star formation stops inside a galaxy. As a result, the black hole mass and the mass of stars in a galaxy are closely linked. Our results also explain for the first time why the quasar lifetime is such a short phase compared to the life of a galaxy."

THE FATE OF GALAXIES

Galaxies are continuing to evolve. Although the changes are slow, the Milky Way galaxy, for instance, will not stay the same forever.

Gravitational attraction of large galaxies such as the Milky Way and Andromeda galaxies draw smaller galaxies. A news release posted at Physorg.com on January 9, 2007, announced the discovery of at least seven small galaxies orbiting the Milky Way. An international research team, including the Penn State University astronomer Donald Schneider and the Cambridge University astronomer Daniel Zucker, made the finding public at a press conference during a meeting in early 2007. The discovery occurred during the Sloan Digital Sky Survey, a research project conducted in several phases from 2000 to 2008. Researchers who conducted this survey scoured more than 25 percent of the sky, imaging and mapping nearly 1 million galaxies and more than 120,000 quasars.

Small galaxies such as these are generally called dwarf galaxies, and they contain only a few million stars, as opposed to the hundreds of billions of stars in the Milky Way and Andromeda galaxies. In the press release, Schneider said, "These dwarf galaxies have been captured by the gravity of the Milky Way and most eventually will merge with our own galaxy."

Absorption of these dwarf galaxies will probably not change the gigantic Milky Way galaxy very much, but the approach of the Andromeda galaxy will have a significant effect. If the galaxies maintain their collision course, they will meet in a few billion years or so. As the galaxies get close, their strong gravitational fields will begin to distort their shapes, elongating and stretching each other. Although collisions between stars are possible, interstellar distances are generally large enough to avoid many direct impacts, although considerable rearrangement may take place among stars and their neighbors. The black holes in the cores will probably combine. In the end, a gigantic elliptical galaxy may emerge, containing the star systems that once formed two large and separate spiral galaxies.

Whatever the Milky Way galaxy's eventual structure, it will one day run out of gas unless additional collisions bring in a new supply. As Zeilik wrote in *Astronomy—The Evolving Universe,* "If we assume that no new gas is added from outside the Milky Way, stellar evolution points to a day when most stars become corpses. Matter that once made up the interstellar medium will be locked up for good. The galaxy will literally run out of gas; starbirth will halt."

The universe will continue to evolve as well. No one knows if the expansion will keep going, or if the universe will stop expanding in the future and maintain a steady state, or perhaps begin contracting. An important factor is the amount of matter in the universe. If the universe contains enough mass, including the invisible dark matter, then gravitational forces may one day halt the expansion. But the presence of dark energy, which accelerates the expansion, is another contributing factor. (See chapter 6 for a discussion of dark matter and dark energy.) If the universe continues to expand, clusters of galaxies could wind up as isolated and eventually sterile islands in a vast wasteland of space. But a contraction would squeeze the galaxies together, perhaps all the way back into the incredibly dense point from which the universe sprang.

CONCLUSION

To learn more about the formation, evolution, and fate of galaxies, researchers must collect more data to complete their analysis. Models and simulations can be used to test specific ideas and make predictions to help researchers learn the best places to aim their telescopes. But researchers also need improved instrumentation if they are to probe this frontier of space and astronomy much further. Fascinating details of young galaxies, billions of light-years away, seem to lie just beyond the range of present measuring techniques. Frustrated astronomers gain only fuzzy glimpses of the early universe.

Advances in the field of galaxy research await more sensitive instruments. Ground-based telescopes suffer from limits imposed by the atmosphere, which bends light rays and blocks some of the most interesting parts of electromagnetic spectrum. Earth's gravity causes telescopes mirrors and lenses to sag, contorting their shape and reducing their effectiveness. Although researchers use sophisticated techniques to combat atmospheric distortion and the size limits on telescopes, the equipment is expensive and of limited benefit.

Elevating an observatory above the atmosphere and into a "weightless" (actually, free-fall) environment of an orbiting satellite is one way to avoid these problems. Astronomers have made a lot of progress with the aid of the *Hubble Space Telescope* and many other orbiting observatories, such as *GALEX* and the *Chandra X-Ray Observatory.* But the technology of the *Hubble Space Telescope* is relatively old, and the instrument is aging. NASA launched the *Hubble Space Telescope* in 1990, and it will not last much longer.

But a replacement is in the works. For several years, NASA has been making plans for another orbiting telescope, formerly known as the *Next Generation Space Telescope* but since 2002 called the *James Webb Space Telescope (JWST),* named in honor of the administrator who headed NASA from 1961 to 1968, during the important development phase of the Apollo moon landing program. NASA began construction in 2008, and expects a launch date in 2013. *JWST* is an international effort involving the European Space Agency (ESA) and the Canadian Space Agency (CSA) as well as NASA.

The study of young galaxies and the early universe is one of the main goals of the telescope. Mission scientists wish to extend their vision all the way back to the universe's Dark Ages, finding the first light that breaks through, as well as the active young galaxies in the process of forming.

JWST's instruments will primarily operate in the infrared range of the electromagnetic spectrum, although they will also have some ability to detect and focus visible light as well. There are several reasons scientists chose infrared radiation. Electromagnetic waves of this wavelength can probe dusty regions that obscure visible light. Another reason is that radiation from faint, distant objects is redshifted, and better viewed in infrared, which has a lower wavelength than visible light. Astronomers also want to study low-temperature sources, which emit mainly infrared radiation.

But infrared telescopes have special requirements. All warm objects emit infrared radiation, including the telescope itself. (If an object gets hot enough, it will emit visible light, but at room temperature, the emissions are mostly infrared.) This radiation tends to drown out faint signals that astronomers wish to study. To avoid the problem, infrared instruments must be kept at extremely cool temperatures of less than -370°F (-223°C). Space itself is frigid—its temperature reads about -455°F (-270°C). But sunlight would send *JWST*'s temperature soaring as the instrument absorbs this energy, so the observatory will include a giant shield to keep this from happening.

The mirror will have a diameter of 21.3 feet (6.5 m), more than twice as large as the *Hubble Space Telescope*'s main mirror. This mirror is so large that it cannot fit into the rocket that will launch the satellite, so it must be folded and pop open after *JWST* deploys. The sunshade must also be stored in this manner.

With the use of *JWST* and similar advanced instruments, astronomers will soon be able to peer farther into the depths of space than ever before. This means that they will be able to see farther back in time as well. These views will help researchers learn much more about how galaxies formed and evolved.

CHRONOLOGY

1774	The French astronomer Charles Messier (1730–1817) begins to publish a list of notable objects in the sky.

1888	The Danish-Irish John L. E. Dreyer (1852–1926) compiles and publishes the first version of the New General Catalog.
1912	The American astronomer Henrietta Leavitt (1868–1921) discovers the period-luminosity relationship in an important class of stars known as Cepheid variables.
1916	The German-American physicist Albert Einstein (1879–1955) proposed the general theory of relativity.
1920	The astronomers Harlow Shapley (1885–1972) and Heber Curtis (1872–1942) meet and debate the nature of the nebula.
1924	The American astronomer Edwin Hubble (1889–1953) uses Cepheid variables to demonstrate that some of the faint nebulae observed by astronomers are much more distant than would be expected if they are part of the Milky Way galaxy.
1925	The Swedish astronomer Bertil Lindblad (1895–1965) proposes the density wave model to explain the spiral arms of galaxies.
1926	Hubble classifies galaxies based on shape.
1929	Hubble formulates the relation between distance and velocity known as Hubble's law.
1960	The astronomers Allan Sandage and Thomas Matthews identify the first quasar.
1980	The theoretical physicist Alan Guth proposes the inflationary theory of the universe's evolution.
1989	*Cosmic Background Explorer (COBE)* is launched.
1990	The *Hubble Space Telescope* is launched.

1993	Space shuttle astronauts fix a minor flaw in the optics of the *Hubble Space Telescope.*
2001	*Wilkinson Microwave Anisotropy Probe (WMAP)* is launched.
2003	*Galaxy Evolution Explorer (GALEX)* is launched.
2007	Researchers use *GALEX* to find a small number of young galaxies billions of light-years away.
2008	The *James Webb Space Telescope* enters implementation phase after a successful review of the project proposal.

FURTHER RESOURCES
Print and Internet

Bone, Neil. *Deep Sky Observer's Guide.* Buffalo, N.Y.: Firefly Books, 2005. This guide for backyard astronomers explains how to find and view distant objects with binoculars or telescopes.

Carnegie Mellon University. "Simulations Show How Growing Black Holes Regulate Galaxy Formation." News release, February 9, 2005. Available online. URL: http://www.cmu.edu/PR/releases05/050209_ blackhole.html. Accessed July 27, 2009. Researchers use computer simulations to study quasars and the evolution of galaxies.

Hoskin, Michael A. "The 'Great Debate': What Really Happened." *Journal for the History of Astronomy* 7 (1976): 169–182. Available online. URL: http://antwrp.gsfc.nasa.gov/htmltest/gifcity/cs_real.html. Accessed July 27, 2009. This article describes the 1920 debate between Shapley and Curtis concerning the nature of nebulae.

Jet Propulsion Laboratory. "Galaxy Evolution Explorer Celebrates Five Years in Space." News release, April 28, 2008. Available online. URL: http://www.galex.caltech.edu/newsroom/glx2008-02f.html. Accessed July 27, 2009. *GALEX* has found millions of galaxies in its first five years.

———. "Watching Galaxies Grow Old Gracefully." News release, November 14, 2007. Available online. URL: http://www.galex.caltech.edu/newsroom/glx2007-05f.html. Accessed July 27, 2009. *GALEX* scientists find "transition" galaxies in the process of becoming older elliptical galaxies.

Melia, Fulvio. *The Edge of Infinity: Supermassive Black Holes in the Universe.* Cambridge: Cambridge University Press, 2003. This book describes the strides researchers have recently made in finding and understanding black holes.

Physorg.com. "Seven or Eight Dwarf Galaxies Discovered Orbiting the Milky Way." News release, January 9, 2007. Available online. URL: http://www.physorg.com/news87576087.html. Accessed July 27, 2009. Researchers announce the discovery of at least seven small galaxies near the Milky Way galaxy.

ScienceDaily. "A Cosmic Fossil? Brilliant But Fuzzy Galaxy May Be Aftermath of Multi-Galaxy Collision." News release, February 7, 2008. Available online. URL: http://www.sciencedaily.com/releases/2008/02/080205115813.htm. Accessed July 27, 2009. A new image from the *Hubble Space Telescope* suggests past merger activity of NGC 1132.

———. "Astronomers Trace the Evolution of the First Galaxies in the Universe." News release, September 13, 2006. Available online. URL: http://www.sciencedaily.com/releases/2006/09/060913190315.htm. Accessed July 27, 2009. The University of California, Santa Cruz, researchers Rychard Bouwens and Garth Illingworth have found plenty of bright galaxies dating from about a billion years after the big bang but practically none earlier.

———. "Black Holes Lead Galaxy Growth." News release, January 7, 2009. Available online. URL: http://www.sciencedaily.com/releases/2009/01/090106181729.htm. Accessed July 27 2009. A study of young galaxies suggests black holes are relatively larger, compared to the mass of the central bulge, than in older galaxies.

Space Telescope Science Institute. "Scientists Find Faint Objects with Hubble that May Have Ended the Universe's 'Dark Ages.'" News release, January 9, 2003. Available online. URL: http://hubblesite.org/newscenter/archive/releases/2003/05/text/. Accessed July 27, 2009. Researchers announce the discovery of galaxies dating from about a billion years after the big bang.

Waller, William H., and Paul W. Hodge. *Galaxies and the Cosmic Frontier.* Cambridge, Mass.: Harvard University Press, 2003. The authors, both astronomers, discuss what scientists have learned about galaxies with recent ground- and space-based telescopic observations.

Zeilik, Michael. *Astronomy—The Evolving Universe,* 9th ed. Cambridge: Cambridge University Press, 2002. This textbook, designed to accompany an introductory college course in astronomy, discusses the principles of the subject as well as the techniques scientists are currently using.

Web Sites

Chandra X-Ray Observatory. Available online. URL: http://chandra. harvard.edu/. Accessed July 27, 2009. This Web site offers news, images, and information on research done with this orbiting X-ray observatory, which has been involved in black hole studies as well as many other projects.

GALEX: Galaxy Evolution Explorer. Available online. URL: http:// www.galex.caltech.edu/. Accessed July 27, 2009. This Web site provides news and information about this satellite and the discoveries scientists are making with it.

HubbleSite. Available online. URL: http://hubblesite.org/. Accessed July 27, 2009. This Web site offers a wealth of news, images, and discoveries from researchers using the *Hubble Space Telescope.*

James Webb Space Telescope. Available online. URL: http://www.jwst. nasa.gov/. Accessed July 27, 2009. This Web site offers news and information on the proposed replacement for the *Hubble Space Telescope.*

6

THE HIDDEN UNIVERSE: DARK MATTER AND DARK ENERGY

Pioneering astronomers charted the movement of stars and planets and fashioned instruments to observe and magnify the light of these distant objects. But many objects in the universe are hidden from human visual perception. Some of them do not emit or reflect much light in the visible portion of the electromagnetic spectrum or are so distant that little light can be seen. Astronomy blossomed more fully after scientists discovered that visible light revealed only a fraction of the total amount of electromagnetic radiation coming from distant sources. The study of radio waves, infrared, ultraviolet, X-rays, and other frequencies of radiation greatly increased astronomical knowledge, allowing researchers to find and study interstellar gas and strange objects such as pulsars and quasars, as well as providing new perspectives on familiar bodies such as the Sun and planets in the solar system.

But a few objects are not luminous at all. Black holes, for example, possess such strong gravity that not even light can escape their grip. Astronomers cannot directly observe such objects and only become aware of their presence by the effects they exert on other objects. The strong gravitational field of black holes attracts bits of matter, which are accelerated toward the black hole and emit high-frequency radiation, which astronomers can detect and study. Accurate theories such as the general theory of relativity, which predicted the existence of black holes, are a big help in knowing what sort of effects to seek.

Researchers have recently found evidence for two "dark" substances that may not interact with light, or interact only weakly. Dark matter is a nonluminous form of matter that astronomers suspect is responsible for certain gravitational fields, the strength of which appears well beyond what could be generated by visible forms of matter. The other substance, dark energy, seems to exert repellent gravitational forces—the opposite of normal gravitation, which is attractive—that astronomers have proposed to explain certain aspects of the universe's expansion. No theories have emerged to guide researchers, and dark matter and dark energy remain hidden and puzzling. Astronomers do not understand the nature of these substances or their components, yet measurements suggest that most of the universe is composed of these substances.

Dark matter and dark energy are problems that scientists must solve in order to gain a better understanding of the universe's dark side. They are active areas of research and are two of the most important topics at the frontiers of space and astronomy. Dark matter and dark energy are two different concepts, but both are included in this chapter because they both belong to the hidden universe. The first part of the chapter focuses on dark matter, and the latter part on dark energy.

INTRODUCTION

In 1687, the British physicist Sir Isaac Newton (1642–1727) introduced the universal law of gravitation—the gravitational force between two objects is proportional to the product of their masses and inversely proportional to the distance between their centers. This law held firm for centuries, but astronomers noted a few discrepancies such as the orbit of Mercury. Mercury's perihelion—its closest approach to the Sun—precesses or advances by a little bit more than Newton's law of gravitation indicated it should. The discrepancy might have indicated that scientists had not identified all the gravitational sources in the solar system, and people hypothesized the existence of another planet, tentatively named Vulcan, close to the Sun. Astronomers never found Vulcan despite extensive searches, and in 1916, the German-American physicist Albert Einstein (1879–1955) accounted for the discrepancy with a more accurate theory of gravitation, the general theory of relativity, as described in chapter 4.

In 1933, the Swiss-American astronomer Fritz Zwicky (1898–1974) described another gravitational riddle. Zwicky was studying the dynamics of galaxy clusters. Astronomers often study motion in order to infer the amount of mass involved; for example, by measuring the acceleration of a body such as a planet orbiting a massive object such as a star, scientists can determine the mass of the object required to produce this acceleration. Zwicky noticed that the galaxies were moving quickly, suggesting the presence of strong gravitational fields, and therefore a lot of mass to create these fields. But based on the light coming from the stars and galaxies, the mass amounted to less than 100 times what was needed. More detailed measurements in the 1970s confirmed this discrepancy.

Zwicky and other astronomers had found that certain galaxies exerted gravitational forces far in excess of those expected from their luminous matter. One hypothesis to explain this discrepancy was the existence of nonluminous, or dark, matter. The additional mass, although unseen, would account for the extra gravitational forces.

Other evidence for dark matter has emerged. Clouds of hot gas surround large elliptical galaxies (see chapter 5 for more information on elliptical galaxies), and this gas would fly away unless held by the gravitational attraction of the galaxy, the mass of which can be estimated from its luminosity. Measurements of the temperature and pressure of this hot gas indicate how much mass is required to keep the clouds of gas from dispersing into space, and the amount of mass suggested from the luminosity studies is not nearly sufficient. Once again, dark matter seems to be present.

Gravitational lensing provides another clue. Gravitational fields bend the path of light rays because of the curvature of space-time, as given by the general theory of relativity (see the sidebar on page 92). As a consequence, the light from a distant object is bent when it passes through strong gravitational fields such as a galaxy or cluster of galaxies on its way to Earth, as illustrated in the figure. Astronomers call this gravitational lensing, since gravity acts as a lens to bend light. Distortion in the images of certain distant objects allows researchers to calculate the mass of the lenses. The result of these calculations suggests the presence of invisible matter.

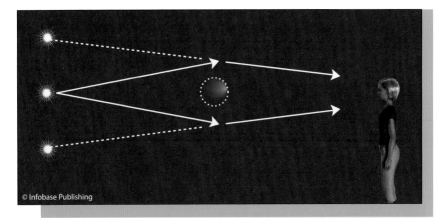

A massive object in the center of the figure bends light rays from a distant star. To an observer's eye, the bent rays seem to be coming from two different objects, located above and below the actual star.

These and other measurements offer no hint as to the nature of the invisible matter. Any object or set of objects could account for the discrepancy as long as this object or set has sufficient mass.

Another option to explain the missing mass is to postulate that it is not required, contrary to theoretical calculations. Perhaps scientists do not understand gravitation as well as they think they do. Consider the wild-goose chase that followed the discovery of the discrepancy in Mercury's orbit, as astronomers scoured the sky for a nonexistent planet when they really should have been looking for a more accurate theory. But no such theory has emerged to explain the dark matter discrepancies, nor is there any reason to look for one except for these discrepancies, which can be neatly explained by the simple assumption of additional, nonluminous mass.

Dark energy is an entirely separate topic. This concept arose because of recent supernova observations and other measurements, as discussed in the following section as well as in the section, "The Accelerating Expansion of the Universe." Dark energy makes its presence known by pushing things apart rather than holding things together, as does dark matter.

But there can be some confusion over the use of the term *dark* for both. Mass and energy are related, as Einstein proposed in 1905, by the formula $E = mc^2$, where E is energy, m is mass, and c represents the speed of light in a vacuum. According to this formula, dark matter has energy associated with it if it has mass, which the observations of Zwicky and others have dictated. But the energy associated with dark matter has nothing to do with dark energy. Dark energy is from a different mold, exhibiting different—even opposing—properties.

In a review on the subject of dark energy in a 2003 issue of *Science,* the Princeton University researchers Jeremiah P. Ostriker and Paul Steinhardt described the differences in dark energy and dark matter. "The only thing dark energy has in common with dark matter is that both components neither emit nor absorb light. On a microscopic scale, they are composed of different constituents. Most important, dark matter, like ordinary matter, is gravitationally self-attractive and clusters with ordinary matter to form galaxies. Dark energy is gravitationally self-repulsive and remains nearly uniformly spread throughout the universe."

To further the study of dark matter and dark energy, researchers need to learn how much of these substances exist in the universe and what composes them. The first question has been easier to answer.

THE AMOUNT OF DARK MATTER AND DARK ENERGY IN THE UNIVERSE

The cosmic microwave background radiation is the remnant of the big bang, in which the universe was created about 14 billion years ago. Coming from all directions, this generally uniform radiation appears as a background in the sky. Astronomers have studied this radiation in order to learn about the universe in its earliest stages. Some small variations or anisotropies have been found in this background, revealed by sensitive instruments on board the satellites *Cosmic Background Explorer (COBE),* which was launched in 1989, and *Wilkinson Microwave Anisotropy Probe (WMAP),* launched in 2001. These variations represent slight differences in temperature.

Scientists have used the anisotropies detected with *WMAP* to construct models of the universe. These models depict the evolution and expansion of the universe and make firm estimates on its density and composition. Although the mathematics and physics are complicated, the models are well fitted to a number of astronomical observations.

As a result of these calculations, scientists can estimate the composition of the energy density of the universe, of which matter is included because of its energy content, as formulated in Einstein's equation. The percentage of the universe's energy that consists of normal, luminous matter is only 5 percent. About 23 percent of the universe is dark matter. The remaining constituent, which at 72 percent comprises the bulk of the universe, appears to be dark energy.

Such a high percentage of nonluminous substances is a little shocking. Yet it should not necessarily be so surprising. Humans evolved to make use of the abundant light available in the form of sunshine—eyes did not evolve to peer into distant parts of the universe. The discovery that the vast majority of the universe's components are dark means that light may be far less important in the universe than it is for human activity.

Researchers participating in COSMOS—Cosmic Evolution Survey—have constructed maps of dark matter. They made these maps by measuring the bending of light from distant sources, which is the same phenomenon that produces gravitational lensing. COSMOS researchers used a variety of telescopes and observatories on the ground and in space, including the *Hubble Space Telescope, Galaxy Evolution Explorer (GALEX), Chandra X-Ray Observatory,* Very Large Array, and others.

In an article published in 2007 in *Seed*, the astronomer and author Phil Plait wrote, "The COSMOS team—over 100 astronomers in a dozen countries—used this formidable array of telescopes to map the positions, distances, and shapes of over 2 million galaxies." The map they produced is impressive. "The result is nothing less than profound: a three-dimensional map millions of light-years across and billions deep, showing the location of trillions of solar masses of invisible ethereal stuff that only decades ago was a complete mystery."

POSSIBLE COMPONENTS OF DARK MATTER

Researchers have indirect evidence that dark matter exists and have mapped its distribution, but no one knows what dark matter is, nor what composes it. Nonluminous objects are by their very nature difficult to find.

The search for dark matter would be easier if researchers had some idea what kind of substance it might be. Perhaps the simplest idea is that dark matter consists of familiar material that is cold, emits little light, and is far removed from bright sources so that it does not reflect much light either. Some of the possibilities are thin clouds of interstellar gas and certain kinds of small stars. For example, brown dwarfs are sometimes called "failed stars" because these objects did not form with quite enough mass to generate the pressure and temperature in their core needed for nuclear reactions to occur. For example, if a star formed from a small interstellar cloud that was less than about 50 to 80 times the size of Jupiter, it would not shine. Although hard to spot, astronomers have found hundreds of brown dwarfs. Neutron stars and other cold, dead stars, well past their shining primes, are also candidates.

Other forms of dark matter could be particles such as *neutrinos.* Neutrinos are highly numerous since they are products of certain nuclear reactions such as those taking place in stars, and they are extremely difficult to detect because they interact only weakly with other matter. Physicists had thought that neutrinos had no rest mass until experiments showed that they can change type—there are three different types of neutrinos, each associated with a specific particle (electron, muon, and tau)—and, according to advanced theories in physics, the ability to change type requires these particles to have mass. No one knows how much mass each neutrino possesses, but it is estimated to be very little.

Black holes would seem ideal candidates for dark matter. These small but massive objects have such strong gravitational fields that light cannot escape, so they are as dark as anything can possibly be.

Although clouds of gas, brown dwarfs, neutrinos, black holes, and other known forms of matter may account for some of the missing mass, there is probably not enough to be responsible for all dark matter. Plenty of neutrinos zip through space, but their mass per particle is tiny. Black holes would seem to be too scarce, and processes that would create a lot of them—large, dying stars—would release considerably more energy and chemical elements with large atomic numbers than has been observed.

Models of the universe, corroborated by *WMAP* and other studies, also cast doubt on the notion that dark matter consists of familiar forms of matter. These studies suggest that the density of protons and neutrons is not high enough to account for all the universe's dark matter. Some researchers are confident in these results. Bernard Sadoulet, a researcher at the University of California, Berkeley, wrote in a 2007 issue of *Science,* "We do know that dark matter is not made of baryons (protons and neutrons), because the baryon density, inferred from the primordial abundance of light elements or the CMBR [cosmic microwave background radiation], is much lower than the total matter density."

If it is not made entirely of familiar particles, what are the constituents of dark matter? No one knows, but the leading candidates include two hypothetical classes of particles—a *weakly interactive massive particle* (WIMP) and an axion. WIMPs zip almost unimpeded through the universe because they have few interactions with ordinary matter and are not luminous, although they have gravitational effects. These particles would be as difficult as or even more difficult to detect than neutrinos, most of which can pass through Earth without pausing. An axion is a particle that physicists have postulated to account for certain issues in quantum mechanics. Its hypothetical properties indicate it would interact only weakly with matter, so it could be a component of dark matter, if it exists. But its predicted mass is small, therefore, many researchers have focused on WIMPs.

DETECTING DARK MATTER PARTICLES

If a particle cannot be measured or observed, it is truly invisible. Some kind of interaction is necessary in order to observe an object—the interaction might be as simple as bouncing a light beam off the object, or it might be complicated and require a lot of energy or a special set of chemicals, but in either case, an interaction is essential. Particles that interact only weakly, such as neutrinos and the hypothetical WIMPs, offer few options for researchers who want to observe them.

Scientists have detected neutrinos with the help of vast underground reservoirs of fluid. Super-Kamiokande, for example, is a neutrino detector consisting of a huge tank that holds about 12,500,000 gallons (47,318,000 L) of pure water, and it is located about 3,280 feet (1,000 m) underground in the Kamioka Mozumi mine in Japan. The underground location blocks everything except weakly interacting particles such as neutrinos, which can plow through rocks without slowing. But every so often, a neutrino hits a water molecule, generating a fast-moving ion that emits a special form of radiation detected by sensitive instruments embedded in the walls of the tank.

Elaborate techniques similar to neutrino detectors will be required to detect WIMPs or axions. Although researchers have already mapped dark matter in the universe, scientists also need a method of closely inspecting these objects in order to study and understand them. In an article written by the journalists Adrian Cho and Richard Stone that appeared in a 2004 issue of *Science,* the Stanford University researcher Edward Baltz said, "It's really going to require that we detect the particles in our galaxy and produce them in the lab, and that we convince ourselves that they are the same thing."

Models of the universe suggest that WIMPs were created during the exceptionally hot, dense conditions that prevailed shortly after the big bang. According to current ideas, WIMPs are moving at about

Soudan Underground Laboratory and the Search for Dark Matter

Iron mining began in Soudan, Minnesota, in 1883 and continued operating for 78 years. In 1965, the Minnesota Department of Natural Resources established the Soudan Underground Mine State Park at the site to preserve the mining heritage. But physicists also became interested in the mine. The dense greenstone in the area has little *radioactivity,* which means that it emits few particles that could interfere with particle detectors and experiments, and it shields the sensitive instruments from other sources on the surface or in space.

As illustrated in the figure, the dark matter detector sits within an "ice box." Liquid helium provides the coolant, which must keep the detector close to the coldest possible temperature—absolute zero, which is about -459.67°F (-273.15°C), or 0 K in the Kelvin scale. At this frigid temperature, the detector can sense low-energy events such as an interaction with a WIMP. The shield provides extra protection against the interference of other particles.

The detector consists of layers of germanium and silicon wafers. In theory, if a WIMP comes along and bumps into one of the atoms in the detector, it will jar a few electrons from one-thousandth the speed of light—which makes them slow movers in terms of particle experiments—and are about 100 to 1,000 times more massive than protons.

One of the leading projects to detect WIMPs is the Cryogenic Dark Matter Search (CDMS). Cryogenic refers to exceptionally cold temperatures. Low temperature is needed in this detector because temperature is a measure of atomic motion—particles move and jiggle around at high speeds when the temperature is high, which would create unwanted noise and mask the signals scientists wish to detect. Researchers

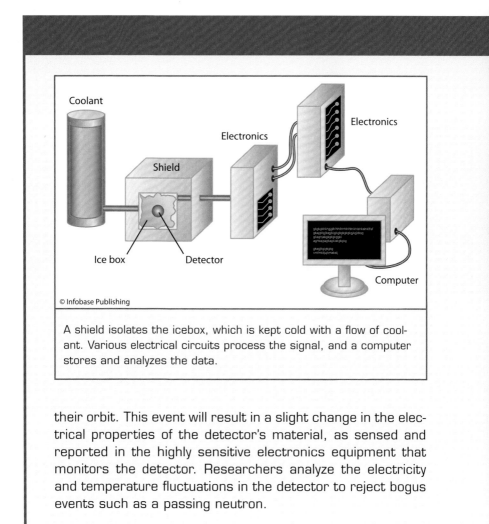

A shield isolates the icebox, which is kept cold with a flow of coolant. Various electrical circuits process the signal, and a computer stores and analyzes the data.

their orbit. This event will result in a slight change in the electrical properties of the detector's material, as sensed and reported in the highly sensitive electronics equipment that monitors the detector. Researchers analyze the electricity and temperature fluctuations in the detector to reject bogus events such as a passing neutron.

conducted the first set of experiments, CDMS I, in a tunnel under the campus of Stanford University in California. No WIMPs were found. Without adequate shielding such as in a laboratory deep underground, too many other particles produced signals that mimicked the quarry.

In 2003, CDMS researchers set up experiments in the Soudan Underground Laboratory and began CDMS II. The Soudan Underground Laboratory is situated about 2,400 feet (730 m) below the surface, at the site of an old iron mine. Other physicists, including some at Fermi National Accelerator Laboratory, also conduct experiments at this location. The sidebar on page 150 provides more information on the laboratory and the dark matter detector that has been installed.

Despite years of operation, CDMS II has not yet found an unambiguous sign of a dark matter particle. But perhaps the detector has failed to reach the level of sensitivity that will be required. Other researchers are planning to build detectors with even greater capabilities.

ZEPLIN is a series of detectors aimed at finding dark matter particles. The second in the series, ZEPLIN-II, is a tank of liquid xenon held at a temperature of -166°F (-110°C), located at the Boulby Underground Laboratory in a mine at North Yorkshire, United Kingdom, 3,600 feet (1,100 m) below the surface. Xenon is a noble gas, but at this cold temperature its phase is liquid. Researchers believe that an interaction between a particle of dark matter and a xenon atom would produce flashes of light called scintillation. (The name ZEPLIN originally stood for "zoned proportional scintillation in liquid noble gases.") Sensitive photodetectors amplify the light, making its detection easier.

In 2007, researchers decided to improve upon the design, creating a ZEPLIN-III detector. A news release posted on April 19, 2007, at ScienceDaily discussed the new plan. "The upgraded ZEPLIN-III, although not significantly bigger than ZEPLIN-II, will be able to achieve a sensitivity that is a factor of 30 better than CDMS, although it should take about two years to reach this level of operation. This factor of 30 is especially important because the theoretical models predict that this is the level of sensitivity needed to have a realistic chance of seeing a signal."

Other instruments that may detect dark matter particles in the future include the *Chandra X-Ray Observatory,* an orbiting X-ray detector that might spot patterns of high-energy radiation coming from collisions or decays of dark matter particles. Another possibility is the Large Hadron Collider, the largest particle accelerator in the world, which began operation in 2008 in Switzerland. The Large Hadron Collider consists of a ring 17 miles (27 km) in circumference, around which protons—which belong to a class of particles known as hadrons—accelerate and then collide, generating a lot of energy and debris consisting of exotic particles. These high-energy collisions may have the potential to create dark matter particles, depending on the mass of these particles. This particle accelerator and other accelerators and observatories are not designed to find dark matter but might encounter such particles in the course of other experiments.

A BETTER UNDERSTANDING OF THE GRAVITATIONAL FORCE

Failure to observe dark matter particles in the laboratory has disappointed researchers. The gravitational effects attributed to this unseen matter are significant, but the absence of direct observation has made some researchers wonder if the hypothesis of dark matter is needlessly complicating the picture. In a 2007 issue of *Science,* the University of Maryland researcher Stacy McGaugh wrote, "This lack of corroboration, combined with the increasing complexity and 'preposterous' nature of a once simple and elegant cosmology, leads one to wonder if perhaps instead gravity is to blame." In other words, perhaps the scientific understanding of gravitational forces is at fault.

In her paper, McGaugh describes a theory known as modified Newtonian dynamics (MOND), which Moti Milgrom of the Weizmann Institute of Science in Israel put forward in 1983. According to McGaugh, "Rather than change the force law at some large length scale, MOND subtly alters it at a tiny acceleration scale, around 10^{-10} m s^{-2} [3.28×10^{-10} feet/second^2]. In systems with gravitational accelerations above this scale (e.g., Earth, the solar system), everything behaves in a Newtonian sense. It is only when accelerations become tiny, as in the outskirts of galaxies, that the modification becomes apparent." In some cases, MOND appears to supplant the need for dark matter to explain the missing mass in the gravitational observations described above. But as McGaugh points out, MOND fails under certain circumstances, and reconciling this theory with the formulas and concepts of gravitation described in the general theory of relativity has been a thorny issue.

Although specific modifications of gravitational laws have not met with wide acceptance, researchers are aware of the limits in certain theories in physics. As discussed in chapter 4, the general theory of relativity is based on concepts that are distinctly different from those of quantum mechanics, which governs the behavior of small particles. These theories appear to be incompatible, and no one understands gravitation on the small scale of particle physics. Scientific understanding of the gravitational force is clearly incomplete.

Recent astronomical observations have suggested that scientific understanding of dark matter's gravitational effects is also far from complete. On August 16, 2007, the Chandra X-Ray Center issued a news release describing a cosmic "train wreck"—a collision of large clusters of galaxies. A research team including Andisheh Mahdavi and Hendrik Hoekstra at the University of Victoria in Canada and their colleagues used the *Chandra X-Ray Observatory* and other telescopes to observe the aftermath of the collision.

Telescopes reveal the distribution of stars in a galaxy, and they can also show astronomers where dark matter is hiding because of its optical distortion. X-ray observatories detect high-energy emissions from objects such as clouds of hot gas. When Mahdavi and his colleagues analyzed data from different instruments, they discovered a core consisting of dark matter and hot gas, but devoid of any bright galaxies. In the news release, Hoekstra said, "It blew us away that it looks like the galaxies are removed from the densest core of dark matter. This would be the first time we've seen such a thing and could be a huge test of our knowledge of how dark matter behaves." Dark matter and galaxies have been intimately associated up until now—recall that Zwicky's gravitational observations of galaxy clusters was the initial suggestion of dark matter's existence—and researchers do not yet understand how even a violent collision could have parted a galaxy from its dark matter.

THE ACCELERATING EXPANSION OF THE UNIVERSE

Dark matter continues to be a mystery, but it is not the only hidden object that researchers seek to study. Another and perhaps deeper mystery lurks in the concept of dark energy. Gravitational forces bound planets orbiting a star, stars within galaxies, and galaxies in clusters. But the universe is expanding, and dark energy may be playing an important role in this expansion.

Cosmological Constant

Astronomers had been viewing stars for centuries before Einstein proposed the general theory of relativity in 1916. In the early 20th century, many scientists considered the Milky Way galaxy as the entire universe. Stars were definitely in motion, but they appeared to be confined within the galaxy. The spreading apart or squeezing together of the universe seemed a strange idea unsupported by any astronomical evidence. When Einstein confronted his theory's prediction of an expanding or contracting universe, he was greatly puzzled.

Throughout Einstein's successful and pioneering career, he did not hesitate to buck conventional wisdom. His theories of time dilation and the quantization of light, although widely accepted today, were groundbreaking when Einstein proposed them. But in this instance, Einstein opted against a bold move. In 1917, Einstein introduced a term into the complicated gravitational field equations to compel a static view of the universe. Walter Isaacson, author of the 2007 biography *Einstein,* wrote, "In his 1917 paper, he was almost apologetic: 'We admittedly had to introduce an extension of the field equations that is not justified by our actual knowledge of gravitation.'" The German term Einstein used was *kosmologische Gleid,* meaning cosmological term or constant.

As evidence for an expanding universe mounted, Einstein willingly abandoned the cosmological constant. Isaacson described the result and gave a quote from the physicist George Gamow, who often consulted with Einstein. "In a new edition of his [Einstein's] popular book on relativity published in 1931, he added an appendix explaining why the term he had pasted into his field equations was, thankfully, no longer necessary. 'When I was discussing cosmological problems with Einstein,' George Gamow later recalled, 'he remarked that the introduction of the cosmological term was the biggest blunder he ever made in his life.'"

Artist's representation of the evolution of the universe, from the big bang (left) through a period of rapid expansion, and then the formation and evolution of the galaxies *(NASA/WMAP Science Team)*

Until the 20th century, people generally viewed the universe as static—the universe was not changing. In 1916, Einstein proposed the general theory of relativity, which describes gravitation as curvature in space-time (the three dimensions of space and one of time). The equations of general relativity are complicated, involving mathematical objects known as tensors, but a simple and unexpected result fell out of these complex formulas. General relativity did not predict a static universe; instead, it predicted a universe that was either expanding or contracting. As discussed in the sidebar on page 155, an embarrassed Einstein proposed to modify the equations with the addition of a term called the *cosmological constant*. This term offset any instabilities, resulting in a prediction of a static universe.

But some scientists did not accept Einstein's constant. In 1922, the Russian mathematician Alexander Friedmann (1888–1925) neglected the term and derived equations based on the general theory of relativity that depicted an expanding universe. The American astronomer Edwin Hubble (1889–1953) found evidence to support the notion of an expanding universe when he noted that distant galaxies are receding at a speed proportional to their distance. As described in chapter 5, this and other evidence has led researchers to develop the present model of the universe. Born about 14 billion years ago in a violent explosion known as the big bang, the universe continues to expand.

To understand the universe, cosmologists must understand the rate at which the universe is expanding. Some theorists hypothesize a period of rapid expansion called inflation early in the universe's history. But following an extremely short inflationary period, the rate of expansion should have been steady or slightly decreasing due to the tug of gravitation. Yet in 1998, researchers performed delicate measurements of distant members of a specific class of supernova that yielded a surprising result—the expansion is accelerating, not slowing down.

The observations involved a certain type of supernova. A Type 1a supernova emits about the same amount of light as others in its class, so the differences in their brightness, as observed from Earth, are due to differences in their distances from the planet. Astronomers calculated these distances based on brightness and measured the redshift, which according to Hubble's law indicates how fast an object is receding. The researchers found that distant supernovae were farther away than they should have been if the expansion had been steady or slowing down. Instead, the expansion is speeding up, or accelerating. Two teams of researchers, one from the Supernova Cosmology Project and the other at the High-Z Supernova Search, made this announcement in 1998. (High-Z refers to objects with a large redshift, implying great distance.)

To explain this acceleration, scientists have proposed the existence of dark energy, which is pushing the universe apart by exerting a sort of "negative pressure" or "repulsive gravity." This is a strange concept, as people are much more familiar with ordinary gravity, but the existence of this kind of energy would account for the supernova observations.

Additional observations using *WMAP* indicate that about 72 percent of the energy density of the universe belongs in the mysterious category of dark energy. But researchers do not yet know anything for certain about the nature of this substance.

THE NATURE OF DARK ENERGY

Dark energy is a simple explanation for the anomalies in the supernova observations and *WMAP* measurements, but just because the existence of this energy would fit the data does not mean it exists. As is the case with dark matter, scientists want independent confirmation. This means that scientists want to observe dark energy and study its properties with some technique that is not linked to the previous observations. Convergence of many independent lines of evidence boosts confidence that a phenomenon is real.

But how do researchers get a handle on this mysterious form of energy? To guide their efforts, researchers usually propose hypotheses. In the case of dark energy, researchers have thought about its possible properties and tried to relate these ideas to objects with which they are already familiar, or measurements they have the ability to perform. Testing these ideas indicates whether researchers are on the right track or not.

A substance as mysterious as dark energy can invoke a considerable amount of imagination. In Adrian Cho's article published in a 2005 issue of *Science,* the University of Chicago theorist Sean Carroll noted, "There are no uninteresting possibilities, which is what makes it so exciting."

Two leading ideas about the nature of dark energy have emerged. One idea is a resurrection of Einstein's cosmological constant. In this scenario, Einstein's introduction of a term in the field equations was not such a "blunder" after all, although he may have gotten the sign of the term wrong. If dark energy is a force that pervades space and exerts pressure that pushes apart the universe, it may well correspond to a cosmological constant.

The other main idea depicts dark energy as a phenomenon that propagates through space similar to electromagnetic radiation. In this scenario, dark energy is a propagating disturbance, the effect of which is to stretch or spread space. This class of theory refers to dark energy as *quintessence,* a term that was once used in ancient and medieval philosophy to describe a fifth element—beyond the traditional four elements of air, earth, fire, and water—that permeates space.

A distinction between these two competing ideas is the strength of dark energy as a function of time. The cosmological constant hypothesis envisions dark energy as relatively unchanging. But in quintessence theories, dark energy is a fluctuating phenomenon.

One way of deciding between these two classes of theory is to make increasingly precise measurements on distant supernovae. In this way, researchers can study the accelerating expansion more carefully and determine whether the magnitude of the phenomenon is changing or staying relatively constant. On February 20, 2004, the Space Telescope Science Institute at Baltimore, Maryland, issued a news release announcing the finding of 42 new supernovae, including some that are among the most distant supernovae ever found. The results suggested that dark energy is not changing very much. This data therefore supports the cosmological constant hypothesis.

But a problem arises when scientists consider the evolution of the universe. Slightly uneven distributions of energy and matter grew into clumps under the attraction of the force of gravitation and later became stars and galaxies. But if dark energy has been a constant force throughout the universe's history, it is difficult to understand how clumps of matter could have collected unless gravitation had been stronger than dark energy at the time. In other words, perhaps dark energy had to be less significant in the early stages of the universe, otherwise stars and galaxies may never have arisen.

A fluctuating energy field, as suggested in quintessence, would explain the varying influence of dark energy. But quintessence has its own set of problems. If dark energy is a variable energy field that exerts repulsive forces, then it is difficult to understand why it seems to be so uniform, pervading all of space. Gravitational attraction results in collections or clumps of matter such as stars and planets forming out of swirling clouds of gas and dust, stars coming together to form galaxies, and galaxies forming clusters and even superclusters (clusters of clusters). Even though dark energy is repulsive, it exerts a force that would seem to result in some sort of uneven distribution.

Another interesting set of questions arise if quintessence is due to some sort of radiation. Radiation such as visible light and other frequencies of electromagnetic radiation show both wave and particle properties; for example, measurement of low levels of light indicate that it consists of pointlike particles called photons, while other measurements demonstrate interference, which is a property of waves. Dark energy should show the same effects, which means particles of dark energy should exist. But if so, particle physicists have seen no sign of them.

The strangeness of the concept of dark energy, combined with the absence of direct observations, has generated some degree of skepticism. A third idea concerning this phenomenon proposes that dark energy is a false premise.

Skepticism is healthy in science. Gullible scientists could easily be led down blind alleys and engage in years or decades of futile experimental and theoretical work. In the case of dark energy, a skeptic might claim that the observations are valid but the conclusion is not. For instance, the universe may appear to be accelerating but this might be because the force of gravitation does not have the same strength over large scales—billions of light-years—that it does on smaller scales, such as star systems and galaxies. But this viewpoint also has problems. If gravitation behaves in this manner, researchers should be able to formulate a set of equations that govern its behavior, and these equations must be consistent with what is already known. But in Cho's 2005 *Science* article, the New York University theorist Gia Dvali noted, "It's very hard to change gravity on large distances without changing it at short distances, too."

Researchers are hard at work, but the nature of dark energy remains one of the most important open questions in space and astronomy. Future researchers must untangle this thicket of competing hypotheses.

FATE OF THE UNIVERSE

Answering the questions about dark energy is critical for several reasons. If dark energy exists as it is currently conceived, it plays an important role in the distribution of matter throughout the universe. Dark energy will also be vital in the future behavior, and the eventual fate, of the universe.

The universe may continue to expand or the expansion may come to a halt and begin to contract. (A steady state in which the universe is not contracting or expanding is unlikely because this situation would not be stable—like a broomstick balanced on one end, a steady universe would eventually tip one direction or the other.) The outcome would be the result of a tug-of-war between the opposing sides of expansion, which pushes the universe apart, and gravitation, which attempts to pull it together.

Before dark energy entered the discussion, cosmologists pondering the fate of the universe studied the density of matter. If the universe contains more matter than a certain critical value—as determined by the general theory of relativity and models of the universe—then the force of gravitation would eventually win and bring the expansion to a halt. A smaller amount of matter would result in an open universe that continues to expand. Although estimates of matter density are difficult to make, astronomical observations indicated that the universe contained an insufficient amount of matter to halt the expansion, even if dark matter is included.

Dark energy introduces another important variable to consider. The opposition to gravitation, provided by the existence of dark energy, would seem to tip the balance even further. Gravity, one might think, would have no chance. In this viewpoint, the universe would not only continue to expand, but it would do so at an increasing rate.

But the situation is not as simple as that. If dark energy represents some kind of cosmological constant, then perhaps the universe will continue to expand at an accelerating rate. But if it exists as a fluctuating field, then its properties are not constant. The acceleration may slow down or even stop. If this is the case, then gravity might still win. The opposite case may also occur—the acceleration may increase, prying the universe apart even faster.

Given the uncertainty in the nature of dark energy, several scenarios are possible. If the repulsive force of dark energy decreases or possibly flips, then the universe may begin contracting some time in the future. This contraction would eventually lead to increasing density and a crowded universe, perhaps eventually reaching a process exactly the opposite of the big bang—the big crunch, in which the universe implodes.

At the other end of the spectrum of possibilities, dark energy may reach a strength in which the universe begins to accelerate at an alarming pace. Such an acceleration may overcome gravity even at small scales, tearing apart galaxies, stars, and possibly even planets. In this scenario, the universe would be ripped apart—the big rip, as some cosmologists have named it.

CONCLUSION

The fate of the universe is billions of years in the future. Events that are so far removed in time, even one of this magnitude, are not a major concern for human civilization, except for scientists and their intellectual curiosity. But the concepts embodied in dark matter and dark energy are critical in the current understanding of the universe. Dark matter and dark energy also affect other branches of science, especially physics.

Experimental laboratories on Earth and ground-based observatories can only do so much. Orbiting telescopes such as the *Hubble Space*

Telescope, Chandra X-Ray Observatory, and others have made important contributions to uncovering the hidden universe, but researchers would also benefit from a project that focuses on dark matter and dark energy. A space-based approach, which has been taken in most of the other topics of this chapter, might lead to important advances in the study of the hidden universe.

In November 2008, NASA and the U.S. Department of Energy (DOE) signed a memorandum of understanding for a cooperative project, the Joint Dark Energy Mission (JDEM). This document defined the roles and responsibilities for each of the participants in this first American space-based mission devoted to the study of dark energy. A news release posted at ScienceDaily on November 25, 2008, announced the memorandum. Dennis Kovar, an official at the Department of Energy's Office of Science for High Energy Physics, said, "DOE and NASA have complementary on-going research into the nature of dark energy and complementary capabilities to build JDEM, so it is wonderful that our agencies have teamed for the implementation of this mission."

The main goal of JDEM is to conduct precision measurements of the universe's expansion over time. Prominent objects to be studied include increasingly distant Type 1a supernovae, as well as phenomena such as gravitational lensing and the clustering and shapes of distant galaxies. To accomplish this goal, current plans include the launch of an optical or infrared orbiting telescope capable of viewing far into space. These instruments, along with deployment and logistic details, are in the design phase. A tentative launch date in 2016 has been proposed.

JDEM researchers aim to address several important questions. A more complete picture of the universe's expansion, especially its time course, should help scientists determine the nature of dark energy. A relatively constant expansion would be consistent with a cosmological constant, in which the dark energy density and its effects have been steady. If dark energy has fluctuated in the past, JDEM researchers need to gauge its evolution in order to understand its nature and role in the universe. The project's scientists would also like to search for signs of any failures in Einstein's general theory of relativity. If this theory, which is the best description of gravitation that scientists have today, does not apply on extremely large-distance scales, then scientists must revisit the basis of the dark energy concept.

Regardless of what this and other future projects discover, the obscure portions of the universe will continue to interest and amaze scientists. The recent finding that much of the universe seems to be hidden was a wake-up call for astronomers and space scientists. Heeding this call, researchers will continue to peer into the depths of space, seeking to unveil these mysteries. A great deal of crucial science remains at this major frontier of space and astronomy.

CHRONOLOGY

1687	The British physicist Sir Isaac Newton (1642–1727) formulates the universal law of gravitation.
1916	The German-American physicist Albert Einstein (1879–1955) proposes the general theory of relativity.
1917	Einstein introduces the cosmological constant in order to stabilize the field equations of the general theory of relativity.
1922	The Russian mathematician Alexander Friedmann (1888–1925) derives equations based on the general theory of relativity depicting an expanding universe.
1929	The American astronomer Edwin Hubble (1889–1953) finds evidence that the universe is expanding.
1931	The Belgian priest and cosmologist Georges Lemaître (1894–1966) proposes the big bang theory.
1933	The Swiss-American astronomer Fritz Zwicky (1898–1974) studies clusters of galaxies and discovers they appear to contain much more mass than their luminosities indicate.
1970s	The Princeton University researcher Jeremiah Ostriker, the Carnegie Institution of Washington researcher Vera Rubin, and others confirm gravitational anomalies that led to the concept of dark matter.

1995	Cryogenic Dark Matter Search (CDMS) begins at Stanford University.
1998	Two teams of researchers—the Supernova Cosmology Project and the High-Z Supernova Search—announce that the expansion of the universe is accelerating.
2001	NASA launches the *Wilkinson Microwave Anisotropy Probe (WMAP)*.
2003	CDMS II begins in the Soudan Underground Laboratory in Minnesota.
2007	Researchers begin upgrading the sensitivity of the ZEPLIN device that aims to detect dark matter particles with a technique based on scintillation in liquid xenon.
	A research team including Andisheh Mahdavi and Hendrik Hoekstra at the University of Victoria in Canada and their colleagues discover a core of dark matter separated from any galaxy.
2008	NASA and DOE sign a memorandum of understanding to establish the Joint Dark Energy Mission.

FURTHER RESOURCES

Print and Internet

Chandra X-Ray Center. "Dark Matter Mystery Deepens in Cosmic 'Train Wreck.'" News release, August 16, 2007. Available online. URL: http://chandra.harvard.edu/press/07_releases/press_081607. html. Accessed July 27, 2009. Researchers using the *Chandra X-Ray Observatory* find dark matter that has been separated from galaxies after a collision of galaxy clusters.

Cho, Adrian. "The Quest for Dark Energy: High Road or Low?" *Science* 309 (September 2, 2005): 1,482–1,483. Cho reviews the arguments for ground- and space-based techniques for studying dark energy.

Hooper, Dan. *Dark Cosmos: In Search of Our Universe's Missing Mass and Energy.* New York: HarperCollins, 2006. Hooper, a scientist who works at Fermi National Accelerator Laboratory, describes why scientists have postulated the existence of dark matter and dark energy, and how they are studying the nature of these mysterious substances.

Isaacson, Walter. *Einstein.* New York: Simon & Schuster, 2007. This engaging biography of Albert Einstein includes materials such as personal letters that have only recently become available.

McGaugh, Stacy. "Seeing Through Dark Matter." *Science* 317 (August 3, 2007): 607–608. McGaugh describes a slight modification of Newton's theory of gravity, which may be able to explain gravitational observations that otherwise seem to require dark matter.

Nicolson, Iain. *Dark Side of the Universe: Dark Matter, Dark Energy, and the Fate of the Cosmos.* Baltimore: Johns Hopkins University Press, 2007. This book offers an accessible introduction to the concepts of dark matter and dark energy.

Ostriker, Jeremiah P., and Paul Steinhardt. "New Light on Dark Matter." *Science* 300 (June 20, 2003): 1,909–1,913. Ostriker and Steinhardt review the current status (up until 2003) of research on dark matter.

Plait, Phil. "Found: Most of the Universe." *Seed.* February 2, 2007. Available online. URL: http://seedmagazine.com/content/article/found_most_of_the_universe/. Accessed July 27, 2009. The astronomer and writer Phil Plait describes the efforts of a team of researchers to map dark matter in the universe.

Sadoulet, Bernard. "Particle Dark Matter in the Universe: At the Brink of Discovery?" *Science* 315 (January 5, 2007): 61–63. Sadoulet reviews efforts to detect particles of dark matter.

ScienceDaily. "NASA and DOE Collaborate on Dark Energy Research." News release, November 25, 2008. Available online. URL: http://www.sciencedaily.com/releases/2008/11/081119171826.htm. Accessed July 27, 2009. Officials announce the signing of a memorandum of understanding to conduct the Joint Dark Energy Mission.

———. "Search for Dark Matter Particles Moves Underground." News release, April 19, 2007. Available online. URL: http://www.sciencedaily.com/releases/2007/04/070419110643.htm. Accessed July 27, 2009. Researchers working on the ZEPLIN project have plans for a new detector with improved sensitivity.

Space Telescope Science Institute. "New Clues about the Nature of Dark Energy: Einstein May Have Been Right After All." News release, February 20, 2004. Available online. URL: http://hubblesite. org/newscenter/archive/releases/2004/12/text/. Accessed July 27, 2009. Researchers using the *Hubble Space Telescope* to study distant supernova acquire data suggesting that dark energy is not changing very much.

Web Sites

Chandra X-Ray Observatory: Dark Matter. Available online. URL: http://chandra.harvard.edu/xray_astro/dark_matter/index.html. Accessed July 27, 2009. This Web site presents the evidence for dark matter and how much the universe contains, along with current theories and attempts to detect dark matter particles.

Cryogenic Dark Matter Search. Available online. URL: http://cdms. berkeley.edu/. Accessed July 27, 2009. CDMS is a project that uses cooled instruments in the attempt to detect dark matter particles and interactions. Their Web site discusses the experiments and presents the latest findings.

National Aeronautics and Space Administration: Joint Dark Energy Mission. Available online. URL: http://jdem.gsfc.nasa.gov/. Accessed July 27, 2009. This Web site provides news and information on JDEM, a space-based attempt to study dark energy.

Final Thoughts

Space exploration is in its early stages, but advances in spacecraft and propulsion methods, along with a determination to succeed, should lead to a more prominent and long-lasting human presence in space. Astronomy will continue to provide essential information and scientific principles to guide these endeavors. Barring a civilization-ending catastrophe, human beings will one day spread throughout the solar system and beyond.

When humans begin space colonization in earnest, Earth will be the origin of a space-faring civilization. But the Milky Way galaxy is home to billions of star systems, at least some of which have planets, and many people wonder if space-faring civilizations have not already arisen in the galaxy or in other parts of the universe. Thousands of planets throughout the galaxy may harbor intelligent life capable of organizing themselves into political and economic societies, which may go on to develop technology, including the ability to travel in space. There is no reason to assume that humans are unique.

But there is also no reason to assume otherwise. Perhaps the development of technological civilization on Earth is unprecedented.

The existence of extraterrestrial (beyond Earth) life and intelligence is an important and fascinating question at the frontier of space and astronomy. Many people have gazed at the stars and wondered if there is another kind of being or life-form, not human but infused with similar emotion and intellect, somewhere among those points of light. But at present there is not enough data to answer this question. One visit from an alien civilization would be sufficient to end all doubt, but as yet no

Wide-field view of a large number of stars and galaxies, some of which may be home to other civilizations *(NASA, ESA, and R. Thompson [University of Arizona])*

convincing evidence of any such visit has ever been found. Claims abound, typically in the form of sightings of unidentified flying objects (UFOs) or tales of encounters or abductions, but no proof has emerged yet to satisfy the scientific community.

If the galaxy contains many space-faring civilizations, it seems strange that Earth has not yet been visited. Although the galaxy is vast, it is billions of years old. If civilizations are plentiful, then some of them must have been around for a long time, unless for some unaccountable reason they have all arisen recently. Billions or even millions of years are enough time for any civilization to spread across the galaxy. But if so, these explorers and colonists left no trace of ever having been on Earth, and they have apparently shied away from this planet in recent times as well, or at the very least avoided skeptics armed with cameras and microphones. But the absence of any conclusive evidence for extraterrestrial civilizations does not prove that none exist.

The subject is rife with speculation, as well as the occasional heated exchange between those who believe in extraterrestrial intelligence and those who do not. Finding a definitive answer will require scientific research. Although the *search for extraterrestrial intelligence* (SETI) is a little bit off the main track (and frequently sullied by sensational claims of UFO devotees), it is a worthwhile frontier of space and astronomy.

One of the first searches began in 1960. The American astronomer Frank Drake (1930–), then at the National Radio Astronomy Observatory (NRAO) in West Virginia, aimed NRAO's 85-foot (26-m) radio telescope at two nearby stars, hoping to detect radio signals that might indicate the presence of intelligent communications. None were found on this early project, called Project Ozma (after a character in L. Frank Baum's fantasy novels). Other similar projects have also failed.

Scientists can only sample a fraction of stars, and they have no way of calculating their chance of success because no one knows the number of civilizations presently residing in the galaxy. These civilizations may number in the thousands, or there may be only one (Earth). In 1961, Drake formulated an equation to estimate the number of active civilizations, N, based on the following factors:

$$N = R^* \cdot f_p \cdot n_e \cdot f_l \cdot f_i \cdot f_c \cdot L$$

where R^* = rate of star formation, f_p = fraction of stars with planets, n_e = number of planets on which life can develop, f_l = fraction of these planets on which life arises, f_i = fraction of life-bearing planets in which intelligent creatures evolve, f_c = fraction of intelligent creatures that develop civilization capable of detection, and L = how long civilizations survive. Scientists can only guess at most of the factors, some of which involve biology, chemistry, and sociology.

But astronomers are beginning to develop tools that may permit them to go beyond guesswork on some of these factors. The *Kepler* Project, described in chapter 1, aims to search 100,000 stars for signs of orbiting planets, especially those with Earth-like properties. In the next five years, the results of this project should indicate how many of this large sample of stars has terrestrial planets. This value will reduce the uncertainty in the Drake equation by at least a little bit.

Researchers are in the process of increasing their odds of success by constructing a radio telescope array devoted to the search for extraterrestrial intelligence. The Allen Telescope Array, a joint project of the SETI Institute and the University of California, Berkeley, began operation in 2007 with 42 of the projected total of 350 antennas. Scientists will use the array, named after the sponsor Paul G. Allen, a businessman and cofounder of Microsoft Corporation, to search a million stars for complex signal patterns that suggest the source is an intelligent community rather than a natural phenomenon.

SETI scientists using the Allen Telescope Array or other sophisticated instruments may one day find unmistakable signs of intelligent life. Although the patterns in these signals may be difficult or even impossible to decipher—and communication at light speed would take years or even centuries to exchange messages—the discovery of civilizations elsewhere in the universe would have a profound philosophical and psychological effect on humans.

But perhaps the most dubious assumption of these searchers is not that such civilizations exist, but that they are enough like human beings to communicate on the same level. Other life-forms may be so alien that even their concept of communication could be fundamentally different. And if this is not enough to prevent communication or even the recognition of such signals, a disparity in the level of technology between human and other civilizations may be too great. Consider that human beings discovered radio waves only a little more than a century ago. Before that time, any attempt of an extraterrestrial civilization to beam a radio message at Earth would have fallen on deaf ears. Advanced civilizations may have developed communication techniques that are completely different from the ones people are now using—possibly as different as the ones modern humans use compared to cavemen.

The reward of a successful SETI project would be great, but the uncertainty of success at this frontier of space and astronomy is also great. Much of the future research in this area depends on advances in many branches of science, which will offer hints as to how many and what sort of beings may be out there. But some of the most important factors are research topics covered in this volume, which will reveal more about the nature of the universe and what it is capable of producing.

GLOSSARY

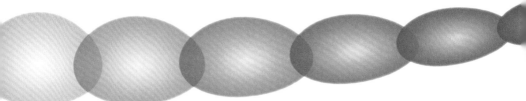

anisotropy   having different properties when measured in different directions or, in other words, along different axes

artificial gravity   production of a force or effect, usually by rotation, that simulates the attraction between masses (gravity)

astronomical unit   the average distance separating Earth and Sun, equal to about 93,000,000 miles (150,000,000 km)

black hole   matter that has collapsed into a tiny volume, exerting a tremendous gravitational field that nothing, not even light, can escape

blueshift   increase in frequency (and corresponding decrease in wavelength) of radiation coming from an approaching object

brown dwarfs   objects with insufficient mass to create enough pressure and temperature to ignite nuclear reactions

Cepheid variables   a class of stars that varies in brightness in a regular manner, often used to gauge astronomical distances because of the relationship between the star's period and luminosity

cosmological constant   a term proposed by Albert Einstein that controls the expansion or contraction of the universe as described in the equations of the general theory of relativity

Doppler effect (or shift)   change in frequency (and a corresponding change in wavelength) of emitted radiation due to relative motion between the source of the radiation and the observer

eccentricity a measure of the departure from a circular shape, with higher values generally representing a more oval or elliptical shape

escape velocity the speed necessary to escape an object's gravitational field

exoplanets *See* **extrasolar planets**

extrasolar planets bodies of considerable size, similar to the planets of the solar system, orbiting another star

extraterrestrial life organisms that do not live on or come from Earth

gamma-ray burst transient source of intense gamma rays

gamma rays high-frequency electromagnetic radiation

generation ship vessel capable of long space voyages, inhabited by successive generations of passengers

graviton a hypothetical particle postulated to be the quantum of gravitational interaction and presumed to have a long lifetime, zero electric charge, and zero rest mass

GRB *See* **gamma-ray burst**

greenhouse gases substances that trap heat by blocking certain forms of electromagnetic radiation

hertz cycles per second of an oscillation or wave

hot Jupiters a class of extrasolar planets that are similar in size to the planet Jupiter but orbit their stars at much closer distances

Hubble constant the value describing the proportional relationship between velocity and distance in Hubble's law

Hubble's law a description of the expansion of the universe in which the velocity of a receding object is proportional to its distance from the observer

interferometry the process of using wave interference to make precise measurements of length and distance

ions electrically charged particles

isotropy having the same or similar properties when measured in different directions or, in other words, along different axes

light-year the distance light travels in a vacuum in a year, equal to 5.88 trillion miles (9.46 km)

luminosity the rate of radiation a body emits

main sequence a band that appears when stars are plotted in a diagram based on spectrum and luminosity, consisting of stars that are active and shining

microgravity a condition in which the acceleration due to the force of gravitation is much weaker than at Earth's surface

momentum the product of an object's mass and velocity, the change of which requires the action of some kind of force

NASA *See* **National Aeronautics and Space Administration**

National Aeronautics and Space Administration the main U.S. agency devoted to space exploration and related sciences and technologies

nebula cloud of interstellar gas or dust

neutrinos elusive particles generated during certain interactions such as the nuclear reactions occurring in stars

neutron star high-density remnant formed when nuclear reactions cease, causing a collapse and compression of matter into tightly packed neutrons

Newton's third law every action has an equal and opposite reaction

orbital period time required for one complete revolution

organic compounds carbon-containing substances normally associated with life or biological processes

neutron star   high-density remnant formed when nuclear reactions cease, causing a collapse and compression of matter into tightly-packed neutrons

pulsar   a source of regular bursts of radio waves, believed to be a rotating neutron star

quantum mechanics   a body of laws and principles in advanced physics that govern the behavior of tiny particles such as atoms and their constituents

quasars   quasi-stellar radio sources—intense emissions of electromagnetic radiation that appear to come from small and distant objects

radial velocity   the velocity of an object along the observer's line of sight, or in other words, how fast the object is moving toward or away from the observer

radioactivity   the emission of particles in the process of nuclear decay or transformation

redshift   decrease in frequency (and corresponding increase in wavelength) in radiation coming from a receding object

search for extraterrestrial intelligence   research aimed at finding signs of intelligent extraterrestrial life, usually by looking for patterned radio signals coming from space

SETI   *See* **search for extraterrestrial intelligence**

solar wind   a high-speed stream of particles that escape from the Sun

space-time   combination of the three coordinates of space and one of time

spectroscopy   the process of analyzing the spectrum of radiation by separating it into its frequency components

terraforming   the shaping or formation of Earth-like (terrestrial) environments

weakly interactive massive particle   a hypothetical object that has mass and exerts gravitational effects but is not luminous

WIMP   *See* **weakly interactive massive particle**

wormhole   tunnels connecting different parts of space and time that may offer a shortcut between two regions of space, or possibly a mechanism permitting time travel

FURTHER RESOURCES

Print and Internet

Andersen, Geoff. *The Telescope: Its History, Technology, and Future.* Princeton, N.J.: Princeton University Press, 2007. This book describes the evolution and potential future development of the fundamental instrument of astronomy.

Bartusiak, Marcia. *The Day We Found the Universe.* New York: Pantheon Books, 2009. The science journalist Marcia Bartusiak recounts the discovery of the expansion of the universe and its cosmological implications.

Christensen, Lars Lindberg, and Robert A. Fosbury. *Hubble: 15 Years of Discovery.* New York: Springer, 2006. This well-illustrated volume documents the many discoveries that astronomers have made using the *Hubble Space Telescope.*

Christianson, Gale. *Edwin Hubble: Mariner of the Nebulae.* Chicago: University of Chicago Press, 1996. This biography tells the story of the astronomer who pioneered the study of the expanding universe.

Couper, Heather, and Nigel Henbest. *The History of Astronomy.* Buffalo, N.Y.: Firefly Books, 2007. Full of anecdotes and brief biographies, this book describes the history of people's fascination with the stars and planets.

Darling, David. *Life Everywhere.* New York: Basic Books, 2002. This book provides a comprehensive discussion of astrobiology—the search for extraterrestrial life, how to find it if it exists, and theories about how it may have evolved.

Florence, Ronald. *The Perfect Machine: Building the Palomar Telescope.* New York: Harper, 1995. What did it take to build the huge 200-inch (5-m) telescope at the Palomar Observatory? This book relates the challenges and triumphs.

Genta, Giancarlo. *Lonely Minds in the Universe: The Search for Extraterrestrial Intelligence.* New York: Springer, 2007. This book discusses the ways in which scientists are searching for signs of extraterrestrial civilizations.

Hawking, Stephen. *The Universe in a Nutshell.* New York: Bantam, 2001. Hawking, a noted physicist at Cambridge University in the United Kingdom, describes Einstein's general theory of relativity, space-time, the evolution of the universe, time travel, and string theory.

Hawking, Stephen, and Leonard Mlodinow. *A Briefer History of Time.* New York: Bantam, 2005. An updated and condensed version of Hawking's *A Brief History of Time,* this book offers an excellent and accessible introduction to advanced physics and cosmology.

Plait, Philip C. *Bad Astronomy: Misconceptions and Misuses Revealed, from Astrology to the Moon Landing "Hoax."* New York: Wiley, 2002. Always on the lookout for scientific fallacies, Plait exposes popular but misguided beliefs concerning space and astronomy.

Tyson, Neil deGrasse. *Death by Black Hole: And Other Cosmic Quandaries.* New York: W. W. Norton, 2007. The author, an astronomer and director of the Hayden Planetarium in New York City, discusses some of the most interesting questions confronting cosmologists.

Tyson, Neil deGrasse, and Donald Goldsmith. *Origins: Fourteen Billion Years of Cosmic Evolution.* New York: W. W. Norton, 2005. A companion to a program aired on the PBS science show *NOVA,* this book summarizes 14 billion years of the universe's history.

Web Sites

European Space Agency (ESA). Available online. URL: http://www.esa. int/. Accessed July 27, 2009. ESA's Web site contains an abundance of information, images, videos, and data concerning space, astronomy, and the agency's long list of projects.

Exploratorium. Available online. URL: http://www.exploratorium.edu/. Accessed July 27, 2009. The Exploratorium, a museum of science, art, and human perception in San Francisco, has a fantastic Web site full of virtual exhibits, articles, and animations, including much of interest to astronomers and space enthusiasts.

How Stuff Works. Available online. URL: http://www.howstuffworks. com/. Accessed July 27, 2009. This Web site hosts a huge number of articles on all aspects of technology and science, including space and astronomy.

National Aeronautics and Space Administration (NASA). Available online. URL: http://www.nasa.gov. Accessed July 27, 2009. NASA's Web site contains a huge amount of information on astronomy, physics, and earth science, and includes news and videos of NASA's many exciting projects.

ScienceDaily. Available online. URL: http://www.sciencedaily.com/. Accessed July 27, 2009. An excellent source for the latest research news, ScienceDaily posts hundreds of articles on all aspects of science. The articles are usually taken from press releases issued by the researcher's institution or by the journal that published the research. Main categories include Matter & Energy, Space & Time, Earth & Climate, and others.

Space.com. Available online. URL: http://www.space.com. Accessed July 27, 2009. This Web site offers news and information on many topics in space exploration and astronomy.

INDEX